Books are to be returned on or before
the last date below.

Revis[e] A2

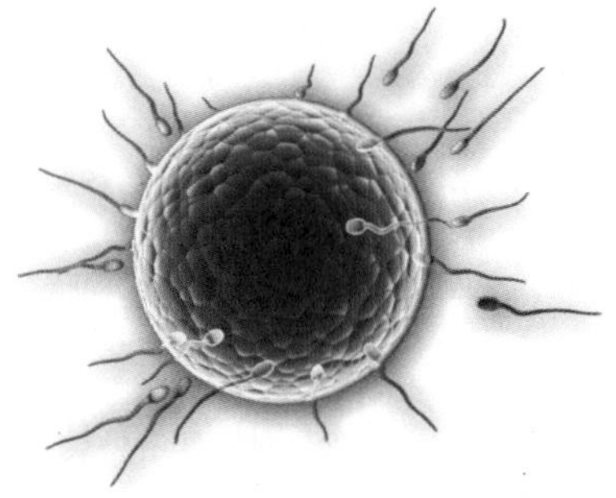

OCR Biology

Joh[...] [...]ysett

058026
THE HENLEY COLLEGE LIBRARY

Contents

THE HENLEY COLLEGE LIBRARY

Chapter 4 Further genetics

Chapter 5 Variation and selection

Chapter 6 Biotechnology and genes

Chapter 7 Ecology and populations

Contents

Specification list

The Specification labels on each page refer directly to the units in the exam board specification, i.e. OCR 4.3.1 *refers to unit 4, module 3, section 1.*

OCR Biology

UNIT	SPECIFICATION TOPIC	CHAPTER REFERENCE	STUDIED IN CLASS	REVISED	PRACTICE QUESTIONS
Unit 4 (M1)	*Communication*	*3.1, 3.2*			
	Nerves	*2.1, 2.2, 2.3*			
	Hormones	*3.1, 3.3*			
Unit 4 (M2)	*Excretion*	*3.3, 3.4*			
Unit 4 (M3)	*Photosynthesis*	*1.2*			
Unit 4 (M4)	*Respiration*	*1.1, 1.3*			
Unit 5 (M1)	*Cellular control*	*4.1, 4.3*			
	Meiosis and variation	*4.2, 5.1, 5.2*			
Unit 5 (M2)	*Cloning in plants and animals*	*6.2*			
	Biotechnology	*6.1*			
	Genomes and gene technology	*6.3*			
Unit 5 (M3)	*Ecosystems*	*8.1, 8.2, 7.1, 7.2*			
	Populations and sustainability	*7.1*			
Unit 5 (M4)	*Plant responses*	*2.6*			
	Animal responses	*2.4, 2.5*			
	Animal behaviour	*2.5, 7.3*			

Examination analysis

Unit 4
1 hour written examination — *A Level – 15%*

Unit 5
1 hour 45 mins written examination — *A Level – 25%*

Unit 6
Internal assessment — *A Level – 10%*

The AS/A2 Level Biology course

AS and A2

All Biology GCE A level courses currently studied are in two parts: AS and A2, with three separate units in each.

Some of the units are assessed by written papers, externally marked by the Awarding Body. Some units involve internal assessment of practical skills (subject to moderation).

Each Awarding Body has a common core of subject content in AS and A2. Beyond the common core material, the Awarding Bodies have included more varied content. This study guide contains the common core material and the additional material that is relevant to the OCR A2 specification.

In using this study guide, some students may have already completed the AS part of the course. Knowledge of AS is assumed in the A2 part of the course. It is therefore important to revisit the AS information when preparing for the A2 examinations.

What are the differences between AS and A2?

There are three main differences:

(i) A2 includes the more **demanding** concepts. (Understanding will be easier if you have completed the AS Biology course as a 'stepping stone'.)

(ii) There is a much greater emphasis on the skills of **application** and **analysis** than in AS. (Using knowledge and understanding acquired from AS is essential.)

(iii) A2 includes a substantial amount of **synoptic** material. (This is the drawing together of knowledge and skills across the modules of AS and A2. Synoptic investigative tasks and questions involving concepts across the specification are included.)

How will you be tested?

Assessment units

OCR A2 Biology comprises three units. The first two units are assessed by examinations.

The third component involves centre assessed practical assessment. This tests practical skills and is marked by your teacher. The marks can be adjusted by moderators appointed by OCR.

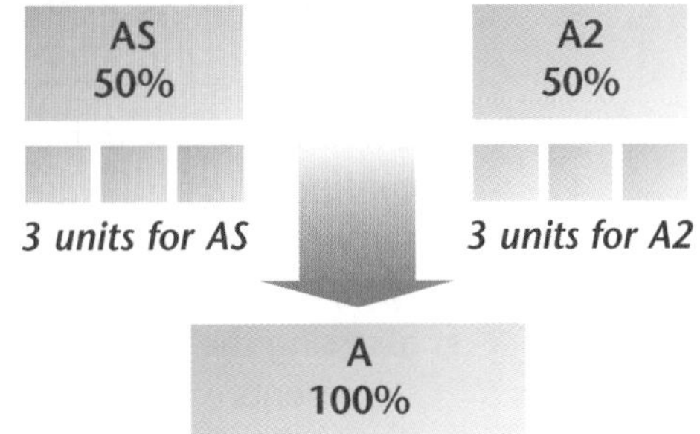

Tests are taken at two specific times of the year, January/February and June. If you are disappointed with a unit result, you can resit each unit any number of times. It can be an advantage to you to take a unit test at the earlier optional time because you can re-sit the test. The best mark from each unit will be credited and the lower marks ignored.

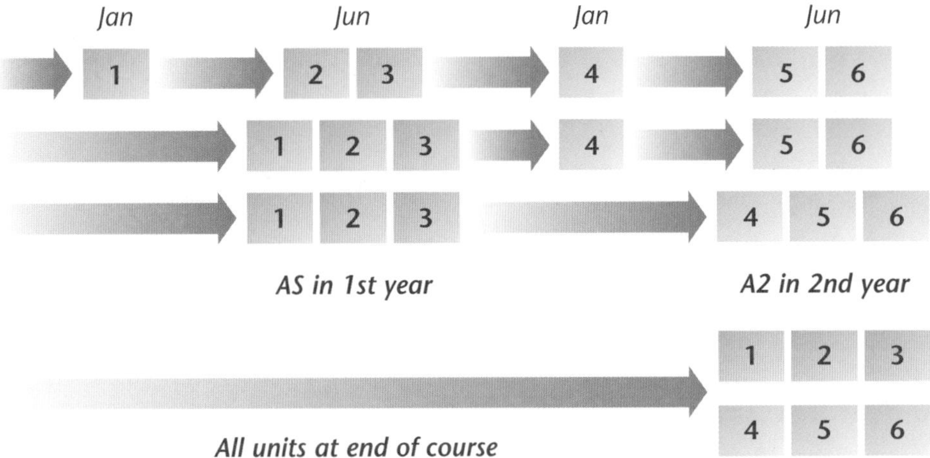

A2 and synoptic assessment

Most students who study A2 have already studied to AS Level. There are three further units to be studied.

Every A Level specification includes synoptic assessment at the end of A2. Synoptic questions draw on the ideas and concepts of earlier units, bringing them together in holistic contexts. Examiners will test your ability to inter-relate topics through the complete course from AS to A2. (See the synoptic chapter page 113).

What skills will I need?

For A2 Biology, you will be tested by assessment objectives: these are the skills and abilities that you should have acquired by studying the course. The assessment objectives are shown below.

Knowledge with understanding

- recall of facts, terminology and relationships
- understanding of principles and concepts
- drawing on existing knowledge to show understanding of the responsible use of biological applications in society
- selecting, organising and presenting information clearly and logically

Application of knowledge and understanding, and evaluation

- explaining and interpreting principles and concepts
- interpreting and translating, from one to another, data presented as continuous prose or in tables, diagrams and graphs
- carrying out relevant calculations
- applying knowledge and understanding to familiar and unfamiliar situations
- assessing the validity of biological information, experiments, inferences and statements

You must also present arguments and ideas clearly and logically, using specialist vocabulary where appropriate. Remember to balance your argument!

Experimental and investigative skills

One of the A2 units tests experimental and investigative skills. The format of this unit is the same as the AS practical unit.

Different types of questions in A2 examinations

Questions in AS and A2 Biology are designed to assess a number of assessment objectives. For the written papers in Biology the main objectives being assessed are:

- recall of facts, terminology and inter-relationships
- understanding of principles and concepts and their social and technological applications and implications
- explanation and interpretation of principles and concepts
- interpreting information given as diagrams, photomicrographs, electron micrographs tables, data, graphs and passages
- application of knowledge and understanding to familiar and unfamiliar situations.

In order to assess these abilities and skills a number of different types of question are used.

In A2 Level Biology unit tests these include short-answer questions and structured questions requiring both short answers and more extended answers, together with free-response and open-ended questions.

Short-answer questions

A short-answer question will normally begin with a brief amount of stimulus material. This may be in the form of a diagram, data or graph. A short-answer question may begin by testing recall. Usually this is followed up by questions which test understanding. Often you will be required to analyse data. Short-answer questions normally have a space for your responses on the printed paper. The number of lines is a guide as to the number of words you will need to answer the question. The number of marks indicated on the right side of the papers shows the number of marks you can score for each question part. Here are some examples. (The answers are shown in blue).

The diagram below shows a gastric pit.

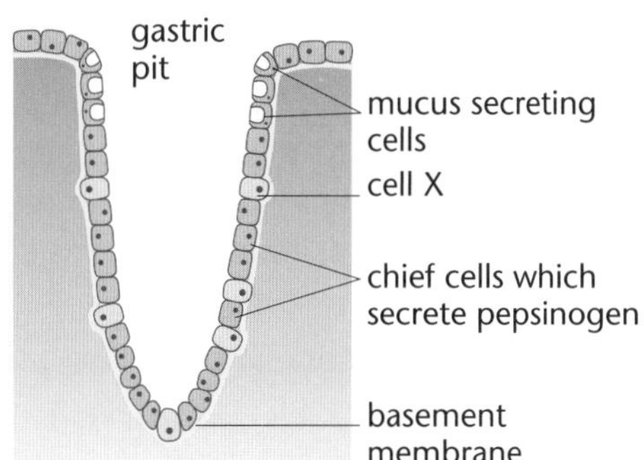

(a) (i) Label cell X (1)

oxyntic cell

(ii) What is secreted by cell X? (1)

hydrochloric acid

(b) (i) Protein enters the stomach. What must take place before the hydrolysis of the protein begins? (2)

Hydrochloric acid acts on pepsinogen, to produce pepsin

(ii) After the protein has been hydrolysed, what is produced? (1)

polypeptides

Structured questions

Structured questions are in several parts. The parts are usually about a common context and they often progress in difficulty as you work through each of the parts. They may start with simple recall, then test understanding of a familiar or unfamiliar situation. If the context seems unfamiliar the material will still be centred around concepts and skills from the Biology specification. (If a student can answer questions about unfamiliar situations then they display understanding rather than simple recall.)

The most difficult part of a structured question is usually at the end. Ascending in difficulty, a question allows a candidate to build in confidence. Right at the end technological and social applications of biological principles give a more demanding challenge. Most of the questions in this book are structured questions. This is the main type of question used in the assessment of both AS and A2 Biology.

The questions set at A2 Level are generally more difficult than those experienced at AS Level. A2 includes a number of higher-level concepts, so can be expected to be more difficult. The key advice given by this author is:

- Give your answers in greater detail:

  Example: Why does blood glucose rise after a period without food?

  Answer: The hormone glucagon is produced ✗ *(This is not enough for credit!)*

  The hormone glucagon is produced which results in glycogen breakdown to glucose. ✓

- Look out for questions with a 'sting in the tail'. A2 structured questions are less straightforward, so look for a 'twist'. This is identified in the example below.

When answering structured questions, do not feel that you have to complete one question before starting the next. Answering a part that you are sure of will build your confidence. If you run out of ideas go on to the next question. This will be more profitable than staying with a very difficult question which slows down progress. You can return at the end when you have more time.

Extended answers

In A2 and AS Biology, questions requiring more extended answers will usually form part of structured questions. They will normally appear at the end of a structured question and will typically have a value of 4 to 10 marks. Longer questions are allocated more lines, so you can use this as a guide as to how many points you need to make in your response. Often for an answer worth 10 marks the mark scheme would have around 12 to 14 creditable answers. You are awarded up to the maximum, 10 marks, in this instance.

Longer extended answers are used to allocate marks for the **quality of communication**.

Candidates are assessed on their ability to use a suitable style of writing, and organise relevant material, both logically and clearly. The use of specialist biological terms in context is also assessed. Spelling, punctuation and grammar are also taken into consideration. Here is a longer-response question.

Question

Urea, glucose and water molecules enter the kidney via the renal artery. Explain what *can* happen to each of these substances.

In this question one mark is available for communication. (Total 10 marks)

Urea, glucose and water molecules can pass out of the blood capillaries in a glomerulus. ✓ This is as a result of ultrafiltration, ✓ as the narrow diameter of the efferent blood vessel cause a pressure build up. ✓

The three substances pass down the proximal tubule. 100% glucose is reabsorbed in the proximal tubule ✓ so is returned to the blood. Carrier proteins on the microvilli aided by mitochondria, actively transport the glucose across the cells. ✓ Around 80% of the water is reabsorbed in the proximal tubule. ✓ Remaining water and urea molecules continue through the loop of Henlé. Urea continues through the distal tubule to the ureter then the bladder. ✓

More water can be reabsorbed with the help of the countercurrent multiplier. ✓ The ascending limb of the loop of Henlé ✓ actively transports Na^+ and Cl^- ions into the medulla. ✓ Water molecules leave the collecting duct by osmosis due to the ions in the medulla. ✓ Cells of the collecting duct are made more permeable to water by the hormone, ADH. ✓ Some water molecules pass into the capillary network and having been successfully reabsorbed. ✓
Some water molecules continue down the ureters and into the bladder. ✓

Communication mark ✓

Remember that mark schemes for extended questions often exceed the question total, but you can only be awarded credit up to the maximum. In response to this question the candidate would be awarded the maximum of 10 marks which included one communication mark. The candidate gave five more creditable responses which were on the mark scheme, but had already scored a maximum. Try to give more detail in your answers to longer questions. This is the key to A2 success.

Stretch and Challenge

Stretch and Challenge is a concept that is applied to the structured questions in Unit 4 and 5 of the exam papers in A2. In principle, it means that sub-questions become progressively harder so as to challenge more able students and help differentiate between A and A* students.

Stretch and Challenge questions are designed to test a variety of different skills and your understanding of the material. They are likely to test your ability to make appropriate connections between different areas and apply your knowledge in unfamiliar contexts (as opposed to basic recall).

Exam technique

Links from AS

A2 builds from the skills and concepts acquired during the AS course. It will help you cope as the A2 concepts ascend in difficulty. The chapters explain the ideas in small steps so that understanding takes place gradually. The final aim, of complete understanding of major topics, is more likely.

Can I use my AS Biology Study Guide for A2?

YES! This will be particularly useful in answering synoptic questions that require direct knowledge of the AS topics.

What are examiners looking for?

Whatever type of question you are answering, it is important to respond in a suitable way. Examiners use instructions to help you to decide the length and depth of your answer. The most common words used are given below, together with a brief description of what each word is asking for.

Define

This requires a formal statement. Some definitions are easy to recall.

Define the term transport.

This is the movement of molecules from where they are in lower concentration to where they are in higher concentration. The process requires energy.

Other definitions are more complex. Where you have problems it is helpful to give an example.

Define the term endemic.

This means that a disease is found regularly in a group of people, district or country.

Use of an example clarifies the meaning. Indicating that malaria is invariably found everywhere in a country confirms understanding.

Explain

This requires a reason. The amount of detail needed is shown by the number of marks allocated.

Explain the difference between resolution and magnification.

Resolution is the ability to be able to distinguish between two points whereas magnification is the number of times an image is bigger than an object itself.

State

This requires a brief answer without any reason.

State one role of blood plasma in a mammal.

Transport of hormones to their target organs.

List

This requires a sequence of points with no explanation.

List the abiotic factors which can affect the rate of photosynthesis in pondweed.

carbon dioxide concentration; amount of light; temperature; pH of water

Describe

This requires a piece of prose which gives key points. Diagrams should be used where possible.

Describe the nervous control of heart rate.

The medulla oblongata ✓ of the brain connects to the sino-atrial node in the right atrium, wall ✓ via the vagus nerve and the sympathetic nerve ✓ the sympathetic nerve speeds up the rate ✓ the vagus nerve slows it down.✓

Discuss

This requires points both for and against, together with a criticism of each point. (**Compare** is a similar command word).

Discuss the advantages and disadvantages of using systemic insecticides in agriculture.

Advantages are that the insecticides kill the pests which reduce yield✓ they enter the sap of the plants so insects which consume sap die✓ the insecticide lasts longer than a contact insecticide, 2 weeks is not uncommon✓

Disadvantages are that insecticide may remain in the product and harm a consumer e.g. humans✓ it may destroy organisms other than the target ✓no insecticide is 100% effective and develops resistant pests.✓

Suggest

This means that there is no single correct answer. Often you are given an unfamiliar situation to analyse. The examiners hope for logical deductions from the data given and that, usually, you apply your knowledge of biological concepts and principles.

The graph shows that the population of lynx decreased in 1980. Suggest reasons for this.

Weather conditions prevented plant growth✓ so the snowshoe hares could not get enough food and their population remained low✓ so the lynx did not have enough hares (prey) to predate upon.✓ The lynx could have had a disease which reduced numbers.✓

Calculate

This requires that you work out a numerical answer. Remember to give the units and to show your working, marks are usually available for a partially correct answer. If you work everything out in stages write down the sequence. Otherwise if you merely give the answer and it is wrong, then the working marks are not available to you.

Calculate the Rf value of spot X. (X is 25 mm from start and solvent front is 100 mm)

$$Rf = \frac{\text{distance moved by spot}}{\text{distance moved by the solvent front}}$$

$$= \frac{25\text{ mm}}{100\text{ mm}} = 0.25$$

Outline

This requires that you give only the main points. The marks allocated will guide you on the number of points which you need to make.

Outline the use of restriction endonuclease in genetic engineering.

The enzyme is used to cut the DNA of the donor cell.✓

It cuts the DNA up like this A T G C C G A T = A T + G C C G A T ✓
T A C G G C T A   T A C G G C   T A

The DNA in a bacterial plasmid is cut with the same restriction endonuclease.✓

The donor DNA will fit onto the sticky ends of the broken plasmid.✓

If a question does not seem to make sense, you may have misread it. Read it again!

Some dos and don'ts

Dos

***Do** answer the question*

No credit can be given for good Biology that is irrelevant to the question.

***Do** use the mark allocation to guide how much you write*

Two marks are awarded for two valid points – writing more will rarely gain more credit and could mean wasted time or even contradicting earlier valid points.

***Do** use diagrams, equations and tables in your responses*

Even in 'essay-style' questions, these offer an excellent way of communicating Biology.

***Do** write legibly*

An examiner cannot give marks if the answer cannot be read.

***Do** write using correct spelling and grammar. Structure longer essays carefully*

Marks are now awarded for the quality of your language in exams.

Don'ts

***Don't** fill up any blank space on a paper*

In structured questions, the number of dotted lines should guide the length of your answer.

If you write too much, you waste time and may not finish the exam paper. You also risk contradicting yourself.

***Don't** write out the question again*

This wastes time. The marks are for the answer!

***Don't** contradict yourself*

The examiner cannot be expected to choose which answer is intended. You could lose a hard-earned mark.

***Don't** spend too much time on a part that you find difficult*

You may not have enough time to complete the exam. You can always return to a difficult calculation if you have time at the end of the exam.

What grade do you want?

Everyone would like to improve their grades but you will only manage this with a lot of hard work and determination. You should have a fair idea of your natural ability and likely grade in Biology and the hints below offer advice on improving that grade.

For a Grade A

You will need to be a very good all-rounder.

- You must go into every exam knowing the work extremely well.
- You must be able to apply your knowledge to new, unfamiliar situations.
- You need to have practised many, many exam questions so that you are ready for the type of question that will appear.

The exams test all areas of the syllabus and any weaknesses in your Biology will be found out. There must be no holes in your knowledge and understanding. For a Grade A, you must be competent in all areas.

For a Grade C

You must have a reasonable grasp of Biology but you may have weaknesses in several areas and you will be unsure of some of the reasons for the Biology.

- Many Grade C candidates are just as good at answering questions as the Grade A students but holes and weaknesses often show up in just some topics.
- To improve, you will need to master your weaknesses and you must prepare thoroughly for the exam. You must become a better all-rounder.

For a Grade E

You cannot afford to miss the easy marks. Even if you find Biology difficult to understand and would be happy with a Grade E, there are plenty of questions in which you can gain marks.

- You must memorise all definitions.
- You must practise exam questions to give yourself confidence that you do know some Biology. In exams, answer the parts of questions that you know first. You must not waste time on the difficult parts. You can always go back to these later.
- The areas of Biology that you find most difficult are going to be hard to score on in exams. Even in the difficult questions, there are still marks to be gained. Show your working in calculations because credit is given for a sound method. You can always gain some marks if you get part of the way towards the solution.

What marks do you need?

The table below shows how your average mark is transferred into a grade.

average	80%	70%	60%	50%	40%
grade	A	B	C	D	E

The A* grade

To achieve an A* grade, you need to achieve a...

- grade A overall (80% or more on uniform mark scale) for the whole A level qualification
- grade A* (90% or more on the uniform mark scale) across your A2 units.

A* grades are awarded for the A level qualification only and not for the AS qualification or individual units.

Four steps to successful revision

Step 1: Understand

- Study the topic to be learned slowly. Make sure you understand the logic or important concepts.
- Mark up the text if necessary – underline, highlight and make notes.
- Re-read each paragraph slowly.

GO TO STEP 2

Step 2: Summarise

- Now make your own revision note summary:
  What is the main idea, theme or concept to be learned?
  What are the main points? How does the logic develop?
  Ask questions: Why? How? What next?
- Use bullet points, mind maps, patterned notes.
- Link ideas with mnemonics, mind maps, crazy stories.
- Note the title and date of the revision notes
  (e.g. Biology: Homeostasis, 3rd March).
- Organise your notes carefully and keep them in a file.

This is now in *short-term memory*. You will forget 80% of it if you do not go to Step 3.
***GO TO STEP 3*, but first take a 10 minute break.**

Step 3: Memorise

- Take 25 minute learning 'bites' with 5 minute breaks.
- After each 5 minute break test yourself:
  Cover the original revision note summary.
  Write down the main points.
  Speak out loud (record on tape).
  Tell someone else.
  Repeat many times.

The material is well on its way to *long-term memory*.
You will forget 40% if you do not do step 4. *GO TO STEP 4*

Step 4: Track/Review

- Create a Revision Diary (one A4 page per day).
- Make a revision plan for the topic, e.g. 1 day later, 1 week later, 1 month later.
- Record your revision in your Revision Diary, e.g.
  Biology: Homeostasis, 3rd March 25 minutes
  Biology: Homeostasis, 5th March 15 minutes
  Biology: Homeostasis, 3rd April 15 minutes
  ... and then at monthly intervals.

Chapter 1

Energy for life

The following topics are covered in this chapter:

- *Metabolism and ATP*
- *Respiration*
- *Autotrophic nutrition*

1.1 Metabolism and ATP

LEARNING SUMMARY

After studying this section you should be able to:

- *understand the principles of metabolic pathways*
- *understand the importance of ATP*

Metabolic pathways

OCR 4.3.1, 4.4.1

Inside a living organism there are many chemical reactions occurring at the same time. They may be occurring in the same place, in different parts of the cell or in different cells. Each reaction is controlled by a different enzyme.

KEY POINT

All the chemical reactions occurring in an organism are called **metabolism**.

Often a number of chemical reactions are linked together. The product of one reaction acts as the substrate for the next reaction. This is called a **metabolic pathway** and each of the chemicals in the pathway are called **intermediates**.

A is the substrate for this pathway, B, C and D are intermediates and E is the product. The enzymes a, b, c and d each control a different step.

$$A \xrightarrow{a} B \xrightarrow{b} C \xrightarrow{c} D \xrightarrow{d} E$$

Metabolic reactions can be classed as one of two types. Reactions that break down complex molecules are called **catabolic reactions** or catabolism. Other reactions build up complex molecules from simple molecules. They are **anabolic reactions** or anabolism.

Anabolic reactions tend to require energy, whereas catabolic reactions release energy. The link between these two types of reactions is a molecule called **ATP**.

Adenosine triphosphate (ATP)

The breakdown of many organic molecules can release large amounts of energy. Similarly, making complex molecules such as proteins requires energy. These reactions must be coupled together. This is achieved by using **adenosine triphosphate (ATP)** molecules.

ATP is a **phosphorylated nucleotide**. (Recall the structure of DNA which consists of nucleotides.) Each nucleotide consists of an organic base, ribose sugar and phosphate group. ATP is a nucleotide with two extra phosphate groups. This is the reason for the term 'phosphorylated nucleotide'.

adenine —— ribose —— phosphate —— phosphate —— phosphate

ATP is produced from adenosine diphosphate and a phosphate group. This requires energy. The energy is trapped in the ATP molecule. An enzyme called ATPase catalyses this reaction.

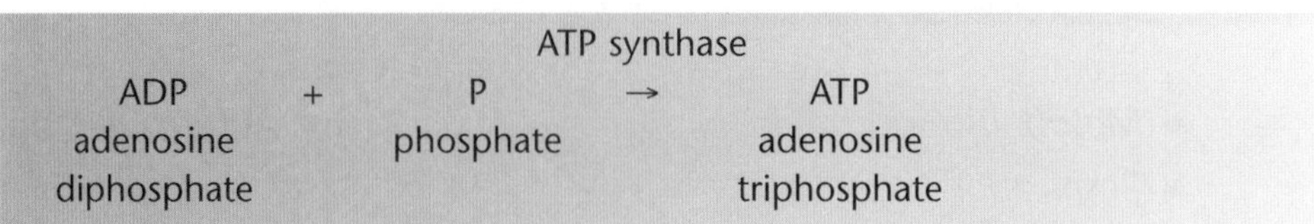

The hydrolysis of the terminal phosphate group liberates the energy. This can then be used in a number of different ways.

ATPase is a hydrolysing enzyme so that a water molecule is needed, but this is not normally shown in the equation.

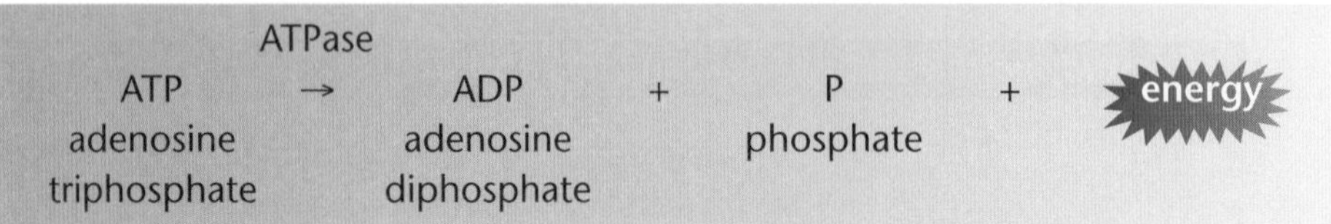

ATP is the cell's energy currency. A cell does not store large amounts of ATP but it uses it to transfer small packets of energy from one set of reactions to another.

KEY POINT

Uses of ATP

- muscle contraction
- active transport
- synthesis of macromolecules
- stimulating the breakdown of substrates to make even more ATP for other uses

1.2 Autotrophic nutrition

After studying this section you should be able to:

- *describe the part played by chloroplasts in photosynthesis*
- *recall and explain the biochemical processes of photosynthesis*
- *relate the properties of chlorophyll to the absorption and action spectra*
- *understand how the law of limiting factors is linked to productivity*
- *understand that glucose can be converted into a number of useful chemicals*

LEARNING SUMMARY

Synthesising food

OCR 4.3.1

Autotrophic nutrition is very important. Autotrophic nutrition means that simple inorganic substances are taken in and used to synthesise organic molecules. Energy is needed to achieve this. In **photo-autotrophic nutrition** light is the energy source. In most instances, the light source is **solar energy**, the process being **photosynthesis**. Carbon dioxide and water are taken in by organisms and used to synthesise glucose, which can be broken down later during respiration to release the energy needed for life. Glucose molecules can be polymerized into starch, a storage substance. By far the greatest energy supply to support food chains and webs is obtained from photo-autotrophic nutrition. Most producers use this nutritional method.

The chloroplast

OCR 4.3.1

Chloroplasts are organelles in plant cells which photosynthesise. In a leaf, they are strategically positioned to absorb the maximum amount of light energy. Most are located in the palisade mesophyll of leaves, but they are also found in both spongy mesophyll and guard cells. There is a greater amount of light entering the upper surface of a leaf so the palisade tissues benefit from a greater chloroplast density.

The diagram below shows the structure of a chloroplast.

Remember that not all light reaching a leaf may hit a chloroplast. Photons of light can be reflected or even absorbed by other parts of the cell. Around 4% of light entering an ecosystem is actually utilised in photosynthesis!

Even when light reaches the green leaf, not all energy is fixed in the carbohydrate product. Just one quarter becomes chemical energy in carbohydrate.

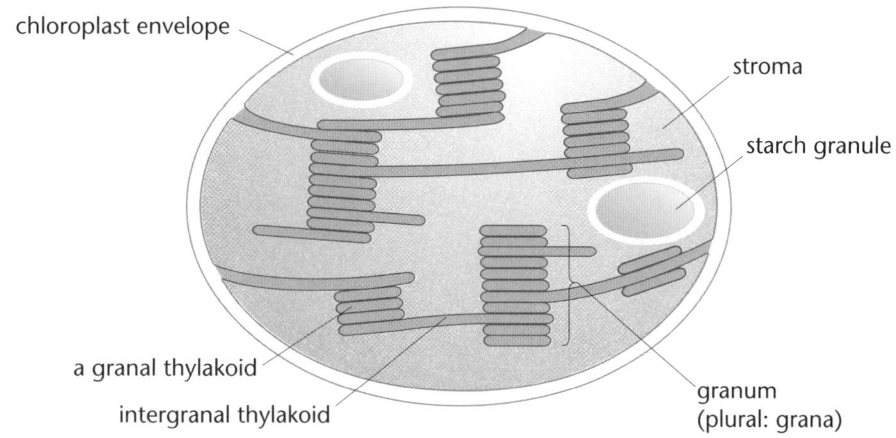

Structure and function

A system of **thylakoid membranes** is located throughout the chloroplast. These are flattened membranous vesicles which are surrounded by a liquid-based matrix, the **stroma**.

Along the thylakoid membranes are key substances:

- chlorophyll molecules
- other pigments
- enzymes
- electron acceptor proteins.

Throughout the chloroplasts, circular thylakoid membranes stack on top of each other to form **grana**. Grana are linked by longer **intergranal thylakoids**. Granal thylakoids and intergranal thylakoids have different pigments and proteins. Each type has a different role in photosynthesis.

The key substances in the thylakoids occur in specific groups comprising pigment, enzyme and electron acceptor proteins. There are two specific groups known as **photosystem I** and **photosystem II**.

Do not be confused by the photosystems. They are groups of chemicals which harness light and pass on energy. Remember this information to understand the biochemistry of photosynthesis.

The photosystems

Each photosystem contains a large number of chlorophyll molecules. As light energy is received at the chlorophyll, electrons from the chlorophyll are boosted to a higher level and energy is passed to pigment molecules known as the **reaction centre**.

KEY POINT

The reaction centre of photosystem I absorbs energy of wavelength 700 nanometres. The reaction centre of photosystem II absorbs energy of wavelength 680–690 nanometres. In this way, light of different wavelengths can be absorbed.

The process of photosynthesis

OCR 4.3.1

The process of photosynthesis is summarised by the flow diagram below.

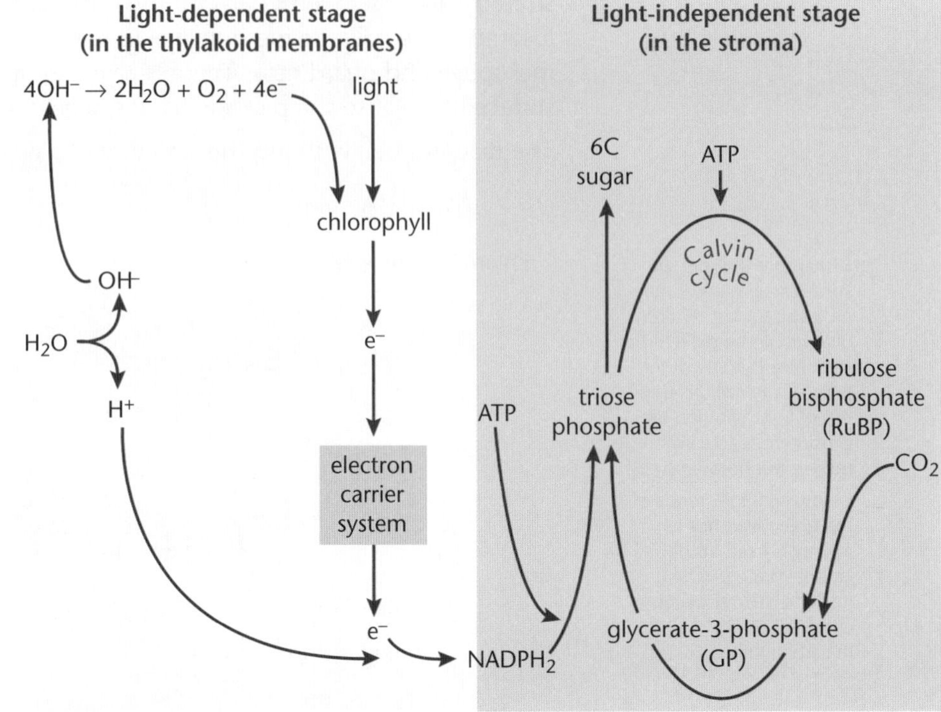

In examinations, look out for parts of this diagram. There may be a few empty boxes where a key substance is missing. Will you be able to recall it?

- Photosynthesis harnesses solar energy.
- Photosynthesis involves light-dependent and light-independent reactions.
- Photosynthesis results in the flow of energy through an ecosystem.

Light-dependent reaction

- Light energy results in the excitation of electrons in the **chlorophyll**.
- These electrons are passed along a series of electron acceptors in the thylakoid membranes, collectively known as the **electron carrier system**.
- Energy from excited electrons funds the production of **ATP** (adenosine triphosphate).
- The final electron acceptor is **$NADP^+$**.
- Electron loss from chlorophyll causes the splitting of water (photolysis):

$$H_2O \rightarrow H^+ + OH^- \quad \text{then} \quad 4OH^- \rightarrow 2H_2O + O_2 + 4e^-$$

- Oxygen is produced, water to re-use, and electrons stream back to replace those lost in the chlorophyll.
- Hydrogen ions (H^+) from photolysis, together with $NADP^+$ form **$NADPH_2$**.

No ATP and $NADPH_2$ in a chloroplast would result in no glucose being made. Once supplies of ATP and $NADPH_2$ are exhausted then photosynthesis is ended. In examinations look out for the 'lights out' questions where the light-independent reaction continues for a while until stores of ATP, $NADPH_2$ and GP are used up. These questions are likely to be graph based.

Light-independent reaction

- Two useful substances are produced by the light-dependent stage, ATP and **$NADPH_2$**. These are needed to drive the light-independent stage.
- They react with glycerate-3-phosphate (GP) to produce a triose sugar – **triose phosphate**.
- Triose phosphate is used ***either*** to produce a 6C sugar ***or*** to form **ribulose bisphosphate** (RuBP).
- The conversion of triose phosphate (3C) to RuBP occurs in the Calvin cycle and utilises ATP, which supplies the energy required.
- A RuBP molecule (5C) together with a carbon dioxide molecule (1C) forms two GP molecules (2 × 3C) to complete the Calvin cycle.
- The GP is then available to react with ATP and $NADPH_2$ to synthesise more triose sugar or RuBP.

How do the photosystems contribute to photosynthesis?

This can be explained in terms of the **Z scheme** shown below.

The **Z scheme**, so called because the paths of electrons shown in the diagram are in a 'Z' shape.

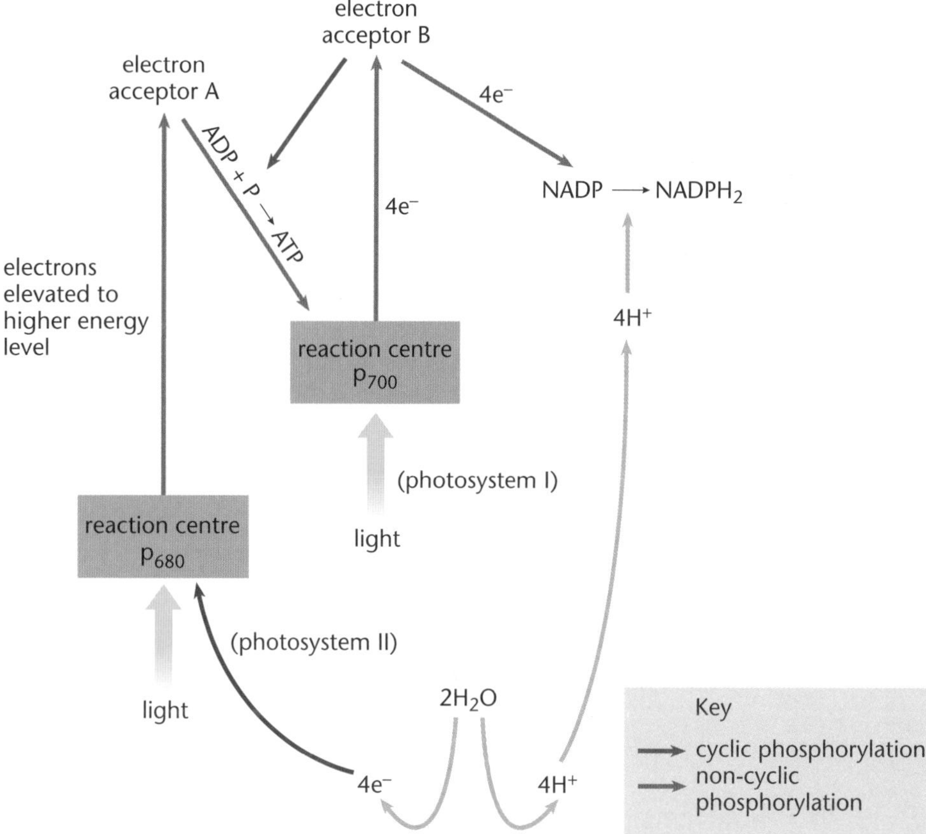

Non-cyclic photophosphorylation

- Light reaches the chlorophyll of both photosystems (P_{680} and P_{700}) which results in the excitation of electrons.
- Electron acceptors receive these electrons (**accepting** electrons is **reduction**).
- P_{680} and P_{700} have become oxidised (**loss** of electrons is **oxidation**).
- P_{680} receives electrons from the **lysis** (splitting) of water molecules and becomes neutral again (referred to as 'hydro lysis').
- Lysis of water molecules releases oxygen which is given off.
- Electrons are elevated to a higher energy level by P_{680} to electron acceptor A and are passed along a series of electron carriers to P_{700}.
- Passage along the electron carrier system funds the production of ATP.
- The electrons pass along a further chain of electron carriers to NADP, which becomes reduced, and at the same time this combines with H^+ ions to form $NADPH_2$.

After analysing this information you will be aware that in cyclic photophosphorylation P_{700} donates electrons then some are recycled back, hence 'cyclic'. In non-cyclic photophosphorylation P_{680} electrons ultimately reach NADP never to return! Neutrality of the chlorophyll of P_{680} is achieved utilising electrons donated from the splitting of water. Different electron sources hence non-cyclic.

Cyclic photophosphorylation

- Electrons from acceptor B move along an electron carrier chain to P_{700}.
- Electron passage along the electron carrier system funds the production of ATP.

Photosynthetic pigments

OCR 4.3.1

Chlorophyll *a* is the only photosynthetic pigment found in all green plants.

The role of photosynthetic pigments is to absorb light energy.

Chlorophyll is not just one substance. There are several different chlorophylls, e.g. chlorophyll *a* and chlorophyll *b*.

- Each is a molecule which has a **hydrophilic head** and **hydrophobic tail**.
- The head always contains a **magnesium** ion and plays a key part in the absorbing or harvesting of light.
- The hydrophobic tail anchors to the thylakoid membrane.

As well as the chlorophylls, there are other **accessory pigments**, e.g. carotenoids which also absorb light energy. There are a range of photosynthetic pigments found in different species.

The graphs below show the specific wavelengths of light which are absorbed by a range of pigments. The data for the **absorption spectrum** was collected by measuring the absorption of a range of different wavelengths of light by a solution of each pigment, chlorophyll *a*, chlorophyll *b*, and carotenoids, **separately**. Following this, plants were illuminated at each wavelength of light, in turn, to investigate the amount of photosynthesis achieved at each wavelength. This data is shown in the **action spectrum**.

The action spectrum shows the actual wavelengths which are used in photosynthesis.

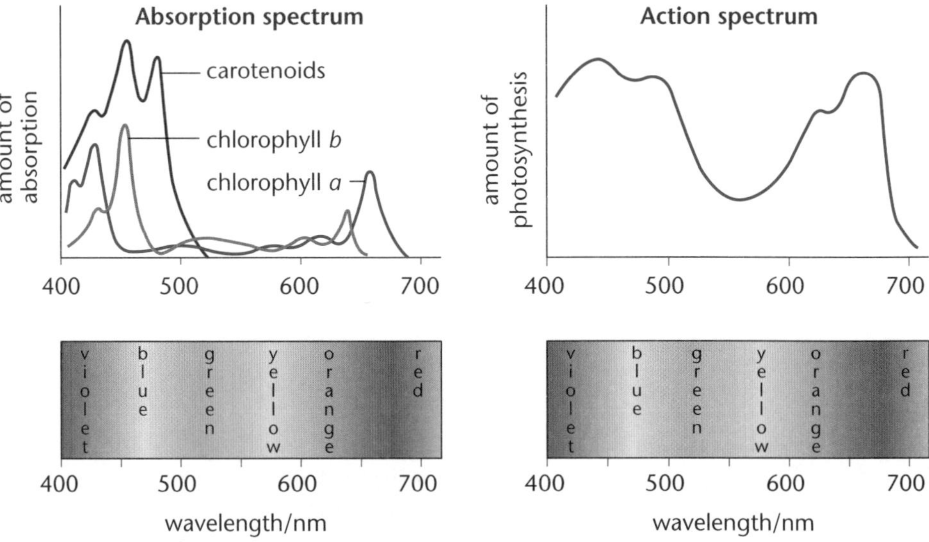

What can be learned from the graphs?

- Blue and red light are absorbed more, and so are key wavelengths for photosynthesis.
- Different pigments have different light absorptive properties.
- Groups of pigments in a chloroplast are therefore much better than just one as more energy can be harnessed for photosynthesis.
- The green part of the spectrum is not absorbed well; no wonder the plants look green as the light is reflected!

Which factors affect photosynthesis?

OCR 4.3.1

If any process is to take place, then correct components and conditions are required. In the case of photosynthesis these are:

- light
- water
- carbon dioxide
- suitable temperature.

Additionally, it is most important that the chloroplasts have been able to develop their photosynthetic pigments in the thylakoid membranes. Without an adequate supply of magnesium and iron, a plant suffers from **chlorosis** due to chlorophyll not developing. The leaf colour becomes yellow-green and photosynthesis is reduced.

Limiting factors

If a component is in low supply then productivity is prevented from reaching maximum. In photosynthesis, **carbon dioxide** is a key limiting factor. The usual atmospheric level of carbon dioxide is 0.04%. In perfect conditions of water availability, light and temperature, this low carbon dioxide level holds back the photosynthetic potential.

Clearly **light energy** is vital to the process of photosynthesis. It is severely limiting at times of partial light conditions, e.g. dawn or dusk.

Water is vital as a photosynthetic component. It is used in many other processes and has a lesser effect as a limiting factor of photosynthesis. In times of water shortage, a plant suffers from a range of problems associated with other processes before a major effect is observed in photosynthesis.

A range of enzymes are involved in photosynthesis; therefore the process has an optimum **temperature** above and below which the rate reduces (so the temperature of the plant's environment can be limiting).

The rate of photosynthesis is limited by light intensity from points A to B. After this, a maximum rate is achieved – the graph levels off.

The rate of photosynthesis is limited by light intensity until each graph levels off. The 30°C graph shows that at 20°C, the temperature was also a limiting factor.

The lower level of CO_2 is also a limiting factor here. The fact that it holds back the process is shown by comparing both graph lines.

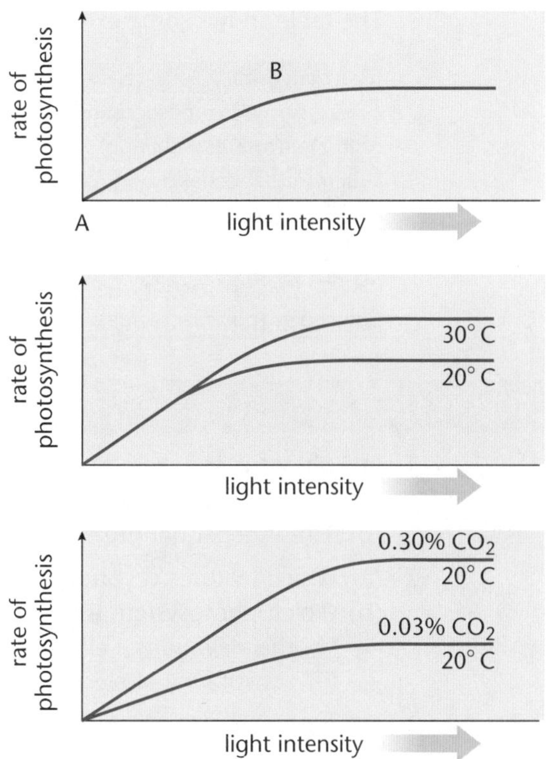

Compensation point

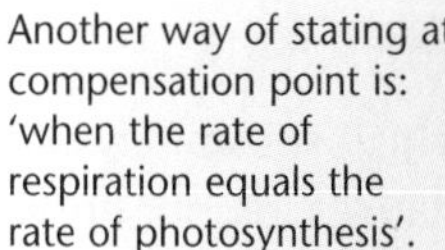

It is usual for a plant growing outside in warm conditions to have **two** compensation points every day.

Photosynthesis utilises carbon dioxide whereas respiration results in its production. At night time during darkness, a plant respires and gives out carbon dioxide. Photosynthesis only commences when light becomes available at dawn, if all other conditions are met. At one point, the amount of carbon dioxide released by respiration is totally re-used in photosynthesis. This is the **compensation point**.

Beyond this compensation point, the plant may increasingly photosynthesise as conditions of temperature and light improve. The plant at this stage still respires producing carbon dioxide in its cells and all of this carbon dioxide is utilised. However, much more carbon dioxide is needed which diffuses in from the air.

In the evening when dusk arrives, a point is reached when the rate of photosynthesis falls due to the decrease in light and the onset of darkness. The amount of carbon dioxide produced at one point is totally utilised in photosynthesis. Another compensation point has arrived.

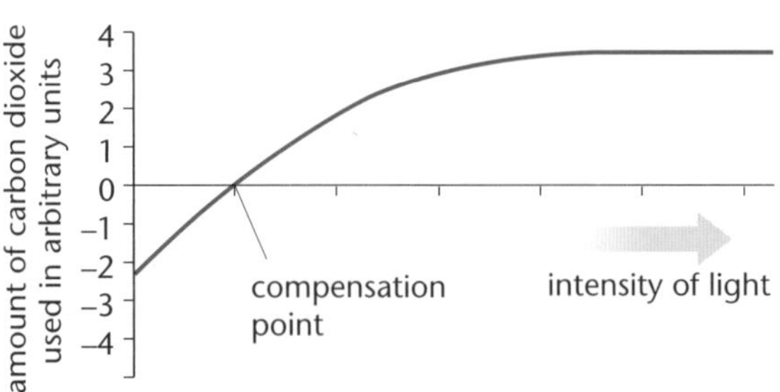

How useful is photosynthesis?

Many more substances are synthesised as a result of photosynthesis. Just a few are highlighted in this section.

Without doubt, photosynthesis is a most important process because it supplies carbohydrates and gives off oxygen. There are many more benefits in that glucose is a 'starter' chemical for the synthesis of many other substances. **Cellulose**, **amino acids** and **lipids** are among the large number of chemicals which can be produced as a result of the initial process of photosynthesis.

The work of the Royal Mint produces the money to run the economy; photosynthesis supplies the **energy currency** for the living world.

The table shows some examples of where and how some carbohydrates are used.

Carbohydrate	*Use*
deoxyribose (monosaccharide)	DNA 'backbone'
glucose (monosaccharide)	leaves, nectar, blood as energy supply
sucrose (disaccharide)	sugar beet as energy store
lactose (disaccharide)	milk as energy supply
cellulose (disaccharide)	protective cover around all plant cells
starch (polysaccharide)	energy store in plant cells
glycogen (polysaccharide)	energy store in muscle and liver

Progress check

1 In a chloroplast, where do the following take place:
(a) light-dependent reaction
(b) light-independent reaction?

2 (a) Which features do photosystems I and II share in a chloroplast?
(b) Which photosystem is responsible for:
(i) the elevation of electrons to their highest level
(ii) acceptance of electrons from the lysis (splitting) of water?

3 (a) Complete the sentence by writing in the correct words.
The compensation point of a plant is when the rate of ....
equals the rate of ....
(b) During a cloudless day in ideal conditions for photosynthesis, how many compensation points does a plant have? Give a reason for your answer.

4 List the three main factors which limit the rate of photosynthesis.

5 During the light-independent stage of photosynthesis, which substances are needed to continue the production of RuBP? Underline the substances in your answer which are directly supplied from the light-dependent stage of photosynthesis.

1 (a) thylakoid membranes (b) stroma.
2 (a) Each photosystem contains a large number of chlorophyll molecules. Light energy is received at the chlorophyll where electrons are boosted to a higher level. Energy is passed to pigment molecules known as the **reaction centre**. The reaction centre of each photosystem absorbs energy (but of different wavelengths!).
(b) (i) photosystem I (ii) photosystem II.
3 (a) respiration; photosynthesis (b) Two. Around dawn and dusk there will come a time when the CO_2 produced as a result of respiration is totally used up in photosynthesis.
4 CO_2; light; temperature.
5 Triose phosphate, NADPH$_2$, ATP and CO_2.

1.3 Respiration

After studying this section you should be able to:

- *recall the structure of mitochondria and relate structure to function*
- *understand that respiration liberates energy from organic molecules*
- *explain the stages of glycolysis and Krebs cycle*
- *explain the stages in the hydrogen carrier system*
- *explain the differences between anaerobic and aerobic respiration*
- *describe different routes which respiratory substrates can take*

LEARNING SUMMARY

The site of respiration

OCR 4.4.1

Respiration is vital to the activities of every living cell. Like photosynthesis it is a complicated metabolic pathway. The aim of respiration is to break down **respiratory substrates** such as glucose to produce **ATP**.

Respiration consists of a number of different stages. These occur in different parts of the cell. Some stages require oxygen and some do not.

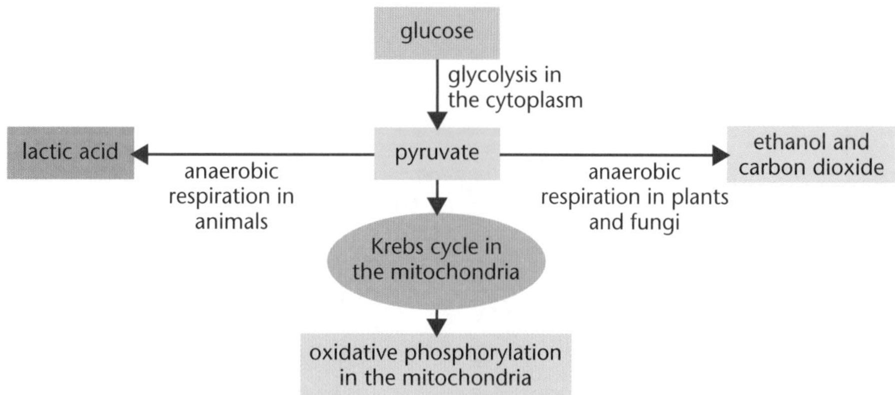

Glycolysis occurs in the cytoplasm of the cell. The pyruvate produced then enters the mitochondria. The Krebs cycle then occurs in the matrix of the mitochondria followed by oxidative phosphorylation which occurs on the inner membrane of the cristae.

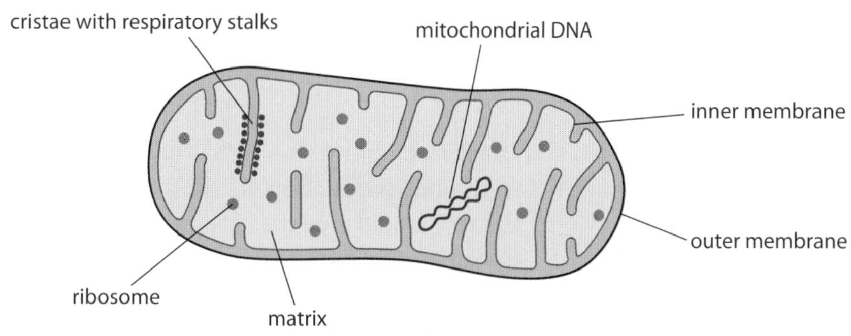

The biochemistry of respiration

OCR 4.4.1

Glycolysis and the Krebs cycle

Both processes produce ATP from substrates but the Krebs cycle produces **many more** ATP molecules than glycolysis. Every stage in each process is catalysed by a specific enzyme. In aerobic respiration, **both** glycolysis and the Krebs cycle are involved, whereas in anaerobic respiration only glycolysis takes place.

The two molecules of ATP are needed to begin the process. Each stage is catalysed by an enzyme, e.g. a decarboxylase removes CO_2 from a molecule.

The flow diagram below shows stages in the breakdown of glucose in glycolysis and the Krebs cycle. The flow diagram shows only the main stages of each process.

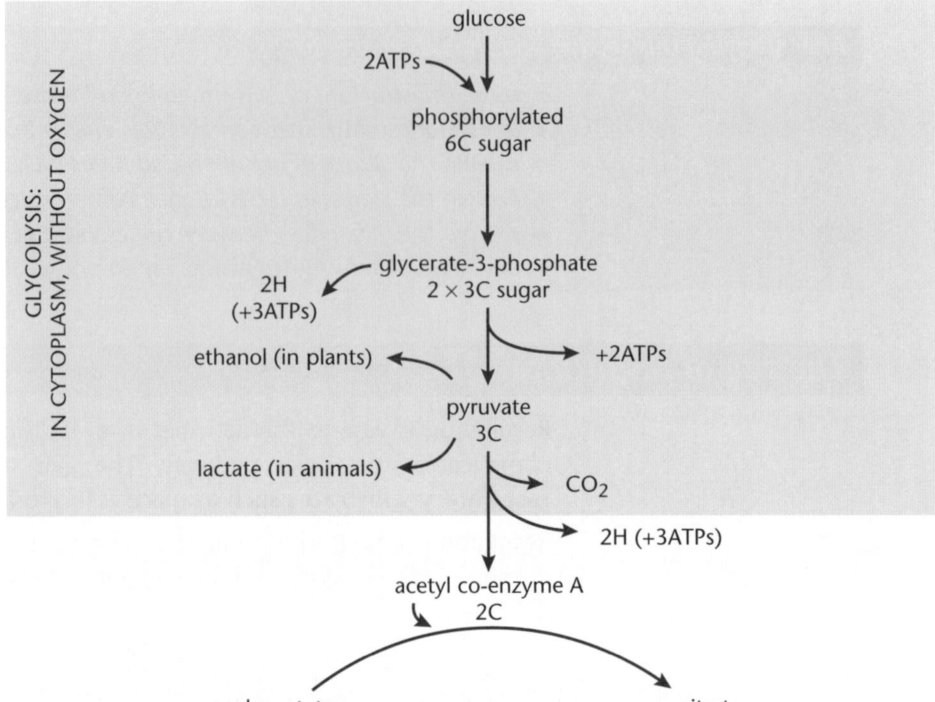

The production of hydrogen atoms during the process can be monitored using DCPIP (dichlorophenol indophenol). It is a hydrogen acceptor and becomes colourless when fully reduced.

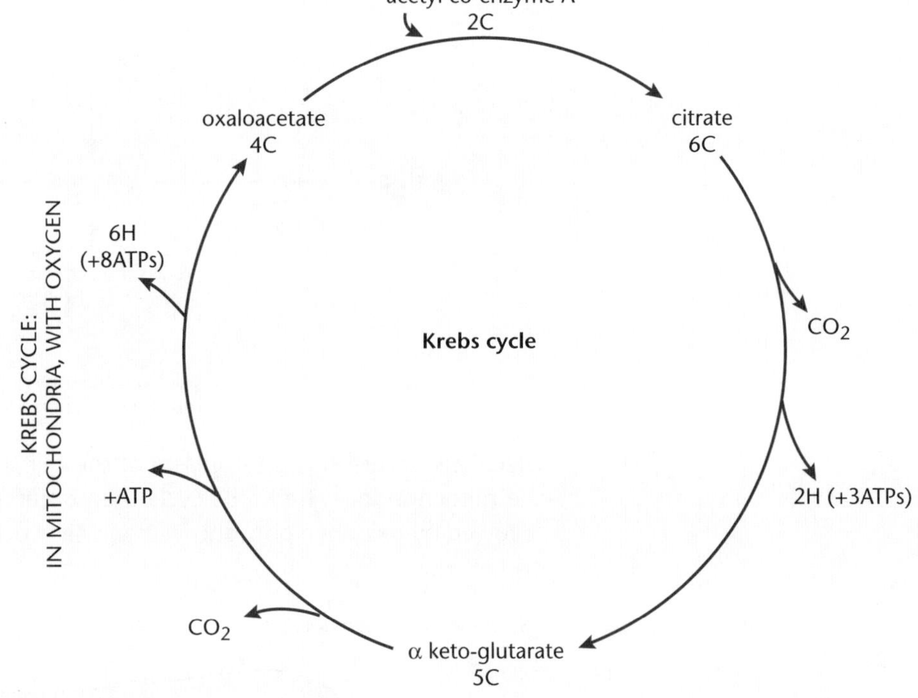

The maximum ATP yield per glucose molecule is:
GLYCOLYSIS 2
KREBS CYCLE 2
OXIDATIVE PHOSPHORYLATION 34
= 38 ATP

The flow diagram shows that glycolysis produces 2 × 2ATP molecules but uses 2ATP so the net production is 2ATP. The Krebs cycle makes 2ATP directly. All the rest of the ATP molecules that are made (shown in brackets) are produced in **oxidative phosphorylation.**

Oxidative phosphorylation

The main feature of this process is the electron carrier or electron transport system. The hydrogen that is given off by glycolysis and the Krebs cycle is picked up by acceptor molecules such as **NAD**. These hydrogen atoms are passed along a series of carriers on the inner membrane of the mitochondrion.

Oxygen is needed at the end of the carrier chain as a hydrogen acceptor. This is why we need oxygen to live. Without it, the generation of ATP along this route would be stopped.

Oxidation	*Reduction*
gain of oxygen	loss of oxygen
loss of hydrogen	gain of hydrogen
loss of electrons	gain of electrons

This is sometimes known as the hydrogen carrier system.

The carrier, NAD, is nicotinamide adenine dinucleotide. Similarly, FAD is flavine adenine dinucleotide.

Hydrogen is not transferred to cytochrome. Instead, the 2H atoms ionise into $2H^+ + 2e^-$. H is passed via an intermediate co-enzyme Q to cytochrome.

Only the electrons are carried via the cytochromes.

e^- is an electron.
H^+ is a hydrogen ion or proton.

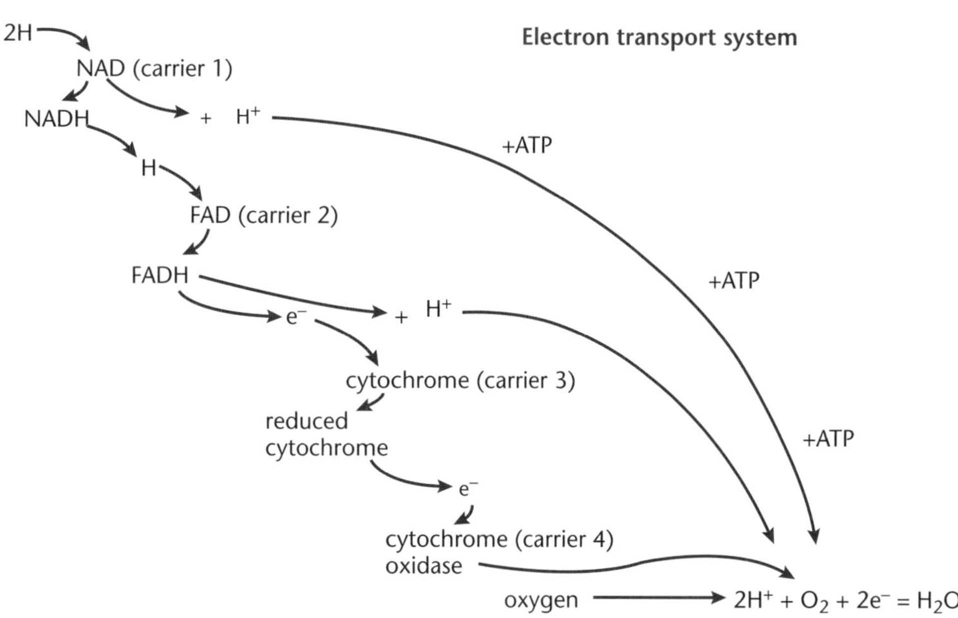

An enzyme can be both an oxidoreductase and a dehydrogenase at the same time!

When oxidation takes place then so does reduction, simultaneously, e.g. $NADH_2$ passes H to FAD. The NAD loses hydrogen and as a result becomes oxidised. FAD gains hydrogen and becomes $FADH_2$, and is therefore reduced. The generic term for an enzyme which catalyses this is **oxidoreductase**. Additionally an enzyme which removes hydrogen from a molecule is a **dehydrogenase**. The result is that three ATPs are produced every time 2H atoms are transported.

The chemiosmotic theory

This theory also explains ATP production in photophosphorylation in the chloroplast. The only difference is that the ions are moved in the opposite direction.

It has now been shown that the carrier molecules are arranged on the membrane of the cristae in a specific way. This means that hydrogen ions are moved out of the matrix and into the space between the two membranes. This sets up a pH gradient. The hydrogen ions can re-enter the matrix through the respiratory stalks. This movement is linked to ATP production and this process is called the **chemiosmotic theory**.

Anaerobic respiration

OCR 4.4.1

If oxygen is in short supply then the final hydrogen acceptor for the hydrogen atoms is missing. This means that oxidative phosphorylation will stop and NAD will not be regenerated. This will result in the Krebs cycle being unable to function.

Ethanol is produced in plants and yeast. Lactate is made in animals.

Glycolysis can continue and produce 2ATP molecules but it would soon run out of NAD as well. A small amount of NAD can be regenerated by converting the pyruvate to lactate or ethanol. This allows glycolysis to continue in the absence of oxygen. This is **anaerobic respiration**.

Anaerobic respiration will make 2ATP molecules from one glucose molecule compared to a possible 38ATP in aerobic respiration.

Other respiratory substrates

OCR 4.4.1

For most cells, glucose is the **preferred respiratory substrate**. This means that a cell will respire glucose before using other substances. Once the supply of glucose and glycogen or starch is exhausted, a cell will then respire other respiratory substrates, such as fats and proteins. These substrates enter the pathway in different places.

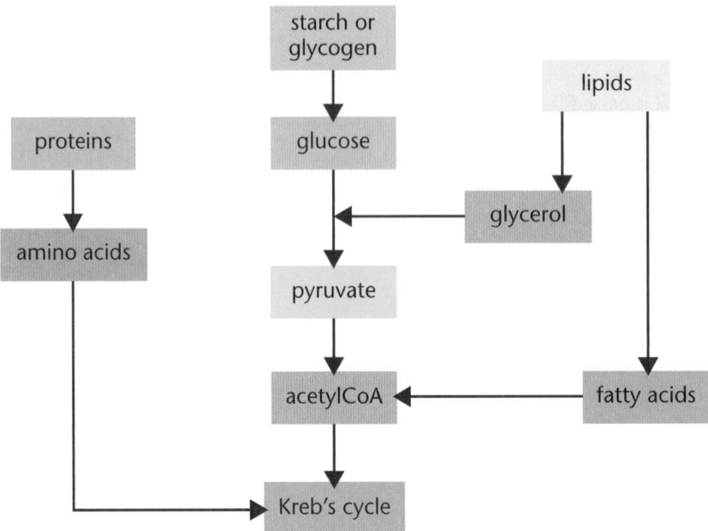

The number of ATP molecules that can be generated from each respiratory substrate depends on the number of hydrogen atoms contained in one mole of the substance. The more hydrogen atoms present, then the more protons that can be released for chemiosmosis. This gives these average energy values for different respiratory substrates:

Respiratory substrate	*Energy value in kJ per gram*
Carbohydrate	15.8
Lipid	39.4
Protein	17.0

Progress check

1 Explain how hydrogen atom production in cells during aerobic respiration results in the release of energy for cell activity.

2 Give **three** similarities between respiration and photosynthesis.

3 (a) Name the **four** carriers in the electron transport system in a mitochondrion. Give them in the correct sequence.

(b) Name the waste product which results from the final stage of the electron transport system.

4 For each of the following statements indicate whether a molecule would be oxidised or reduced.

(a) (i) loss of oxygen
(ii) gain of hydrogen
(iii) loss of electrons

(b) Which type of enzyme enables hydrogen to be transferred from one molecule to another?

1 Used in the electron transport system to produce ATP; 3ATP molecules produced for every 2H atoms produced; ATP → ADP + P + energy released

2 The stages of each process are catalysed by enzymes; both processes involve ATP; respiration involves GP in glycolysis and photosynthesis involves GP in the light-independent stage

3 (a) NAD → FAD → cytochrome → cytochrome oxidase
(b) water

4 (a) (i) reduced (ii) reduced (iii) oxidised
(b) Oxidoreductase

Sample question and model answer

Radioactivity is used to label molecules. They can then be tracked with a Geiger-Müller counter.

In an experiment, pondweed was immersed in water which was saturated with radioactive carbon dioxide ($^{14}CO_2$). It was illuminated for a time so that photosynthesis took place, the light was then switched off. The graph below shows the relative levels of some substances.

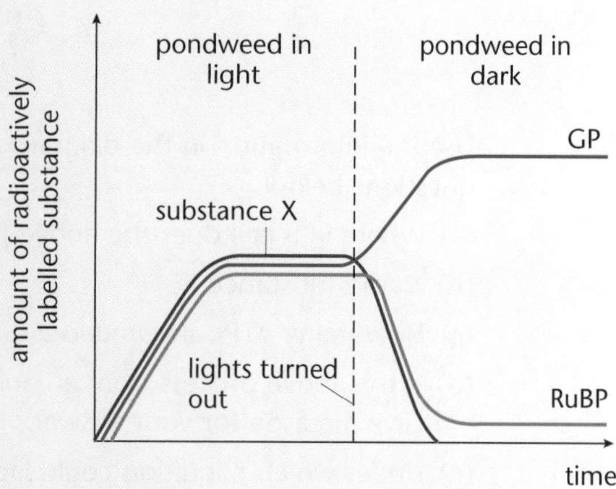

Always be ready to link the rise in one graph line with the dip of another. The relationship holds true here as substance X and RuBP are used up in the production of GP via the Calvin cycle. It is likely that some GP would have been used with substance X to make triose sugar. This is not shown on this graph.

Use the graph and your knowledge to answer the following questions.

(a) (i) Substance X is produced after a substance becomes reduced during the light-dependent stage of photosynthesis. Name substance X. [1]

$NADPH_2$, reduced nicotinamide adenine dinucleotide phosphate

(ii) Explain why substance X cannot be produced without light energy. [3]

- Light energy removes electrons from chlorophyll.
- The electrons are passed along the electron carrier chain.
- The electrons are needed to reduce NADP.

(b) Explain the levels of substance X, GP and RuBP after the lights were turned off. [6]

- It seems that substance X is used to make the other two substances because it becomes used up.
- Supply of substance X cannot be produced without light energy.
- GP is made from RuBP.
- GP levels out because more $NADPH_2$ is needed to make triose sugar or RuBP, the supply being exhausted.
- RuBP levels out at a low level because more $NADPH_2$ is needed to make GP.
- ATP is needed to make RuBP, ATP is needed to make GP.

ATP is not shown on the graph. Always be ready to consider substances involved in a process but not shown. Here it is worth a mark to remember that ATP is needed to continue the light-independent system of photosynthesis.

(c) After the lights were switched off glucose was found to decrease rapidly. Explain this decrease. [1]

- Glucose is used up in respiration to release energy for the cell.

(d) Give the specific sites of each of the following stages of photosynthesis in a chloroplast: [2]

(i) light-dependent stage thylakoid membranes
(ii) light-independent stage. stroma

Practice examination questions

1 The flow diagram below shows stages in the process of glycolysis.

		2ATPs							
glucose	→	phosphorylated	→	GP	→	substance X	→	lactate
6C sugar		6C sugar		glycerate- 3-phosphate (2 × 3C) 2ATPs		3C		

Use the information in the diagram and your knowledge to answer the questions below.

(a) Where in a cell does the above process take place? [1]

(b) Name substance X. [1]

(c) How many ATPs are *produced* during the above process? [1]

(d) Is the above process from an animal or plant?
Give a reason for your answer. [1]

(e) Under which condition could lactate be metabolised? [1]

[Total: 5]

2 The graph shows the relative amount of carbon dioxide taken in or evolved by a plant at different times during a day when the sun rose at 5 a.m.

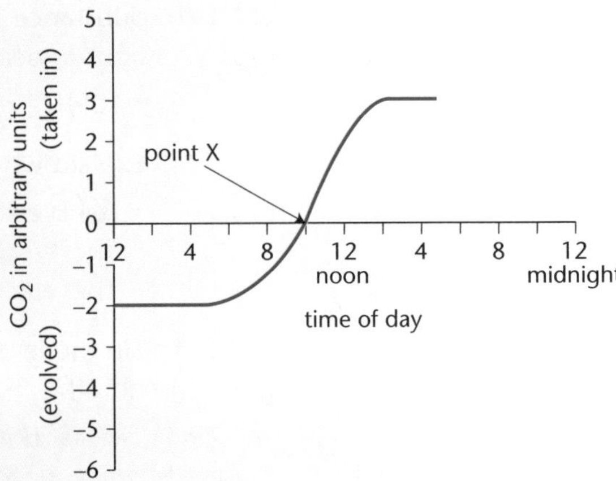

(a) Explain the significance of point X. [2]

(b) What name is given to point X? [1]

(c) Complete the graph between 4.00 p.m. and 12.00 midnight. [2]

[Total: 5]

3 The chlorophyll in a pondweed consisted of several photosynthetic pigments. The graph on the right shows:

(A) the absorption spectrum of the pondweed's chlorophyll *a* measured in arbitrary units

(B) the action spectrum of the same pondweed measured in cm^3 oxygen evolved.

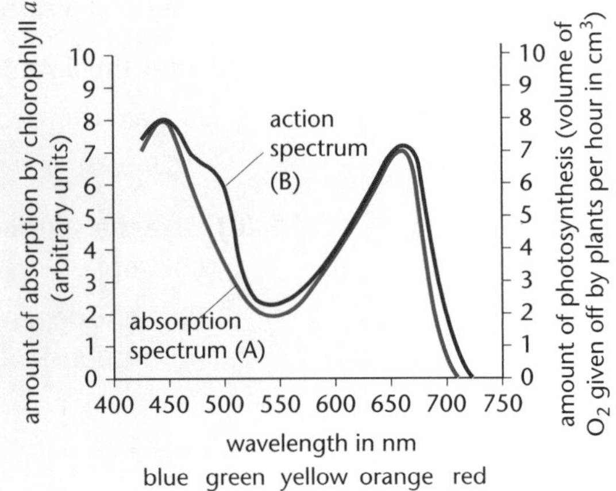

Practice examination questions (continued)

Use the graph and your knowledge to answer the questions.

(a) Explain the similarities and differences between the action and absorption spectra. [2]

(b) Explain the effect of a wavelength of 525 nm on the rate of photosynthesis. [1]

(c) How would the data for the action spectrum have been collected using the pondweed? [1]

[Total: 4]

4 The flow diagram below shows part of the electron carrier system in an animal cell.

$$FADH \rightarrow FAD + H^+ + e^-$$

(a) Where in a cell does this process take place? [1]

(b) From which molecule did FAD receive H to become FADH? [1]

(c) Which molecule receives the electron produced by the breakdown of FADH? [1]

(d) As FADH becomes oxidised a useful substance is produced. Name the substance. [1]

[Total: 4]

5 The graph below shows the effect of increasing light intensity on the rate of photosynthesis of a plant where the concentration of carbon dioxide in the atmosphere was 0.03%.

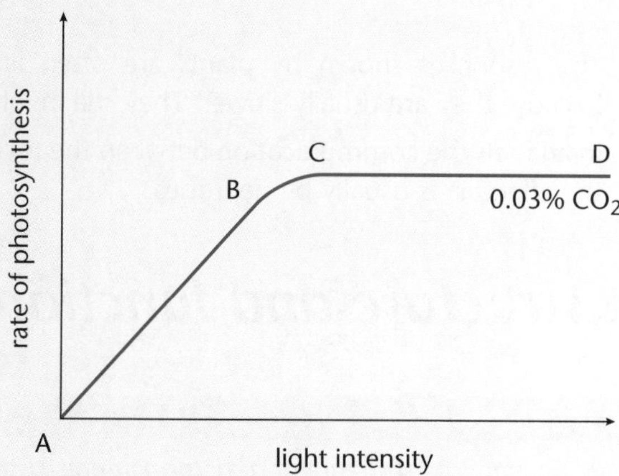

(a) Explain the effect of light intensity on the rate of photosynthesis between the following points on the graph:

(i) A and B

(ii) B and C

(iii) C and D. [3]

(b) Draw the shape of the graph which would result from a CO_2 concentration of 0.3%. [1]

[Total: 4]

Chapter 2

Response to stimuli

The following topics are covered in this chapter:

- *Stages in responding to stimuli*
- *Receptors*
- *Response*
- *Neurone structure and function*
- *Coordination by the CNS*
- *Plant sensitivity*

2.1 Stages in responding to stimuli

LEARNING SUMMARY

After studying this section you should be able to:

- *describe the pathway of events that results in response to stimuli*

The stimulus/response pathway

OCR 4.1.2, 5.4.2

The ability of plants and animals to respond to changes in their external environment is called **sensitivity**. This is a characteristic of all living organisms and is necessary for their survival. Organisms also respond to changes in their internal environment and this is covered in the next chapter.

The events involved in a response follow a similar pattern:

The responses shown by plants are often less obvious than animal responses because they are usually slower. They still involve a similar pathway of events.

In animals the communication between the receptors, the coordinating centre and the effectors is usually by neurones.

2.2 Neurone structure and function

LEARNING SUMMARY

After studying this section you should be able to:

- *describe the structure of a motor neurone, a sensory neurone and a relay neurone.*
- *understand the function of sensory, motor and relay neurones*
- *understand nervous transmission by action potential*
- *describe the mechanisms of synaptic transmission*

The structure and function of neurones

OCR 4.1.2

Neurones are **nerve cells** which help to coordinate the activity of an organism by transmitting **electrical impulses**. Many neurones are usually gathered together, enclosed in connective tissue to form **nerves**.

Important features of neurones.

1 Each has a cell body which contains a nucleus.
2 Each communicates via processes from the cell body.
3 Processes that carry impulses away from the cell body are known as axons.
4 Processes that carry impulses towards the cell body are known as dendrons.
5 All neurones transmit electrical impulses.

KEY POINT

The nervous system consists of a range of different neurones which work in a network through the organs. The diagrams show three types of neurone.

Notice the direction of the impulse and that motor neurones have long axons and short dendrons. This is the other way round for sensory neurones.

Key points from AS

- **The cell surface membrane** *Revise AS pages 49–50*
- **The movement of molecules in and out of cells** *Revise AS pages 51–52*
- **The specialisation of cells** *Revise AS pages 35–36*

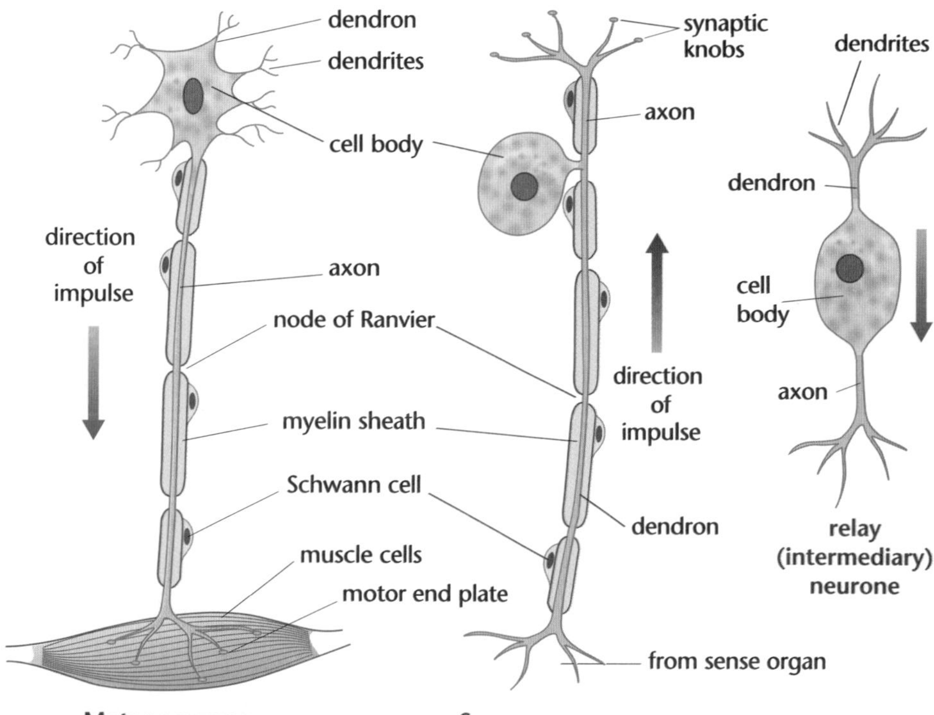

Myelinated neurones

Sensory and motor neurones are examples of myelinated neurones. This enables them to transmit an impulse at a greater velocity. Myelinated neurones have the following characteristics:

- The axon or dendron is insulated by a myelin sheath.
- The myelin sheath is formed by a Schwann cell wrapping around the axon many times. This forms many layers of cell membrane surrounding the axon.
- At intervals there are gaps in the sheath, between each Schwann cell, called nodes of Ranvier.

The myelin sheath is often called a 'fatty' sheath because it is made of many layers of cell membrane which are composed largely of phospholipids.

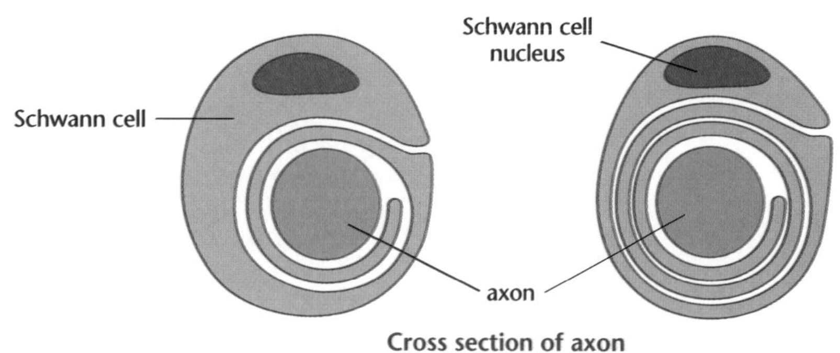

Cross section of axon

What are the roles of the sensory and motor neurones?

There are many similarities between the structure of sensory and motor neurones but they have different functions:

- The sensory neurones transmit impulses towards the central nervous system (CNS) from the receptors.
- The motor neurones transmit impulses from the CNS to effectors, such as muscles, to bring about a response.
- Relay neurones may form connections between sensory and motor neurones in the CNS.

Transmission of an action potential along a neurone

OCR 4.1.2

Neurones can 'transmit an electrical message' along an axon. However, you must never write this in your answers. Instead of nerve impulse, you must now use the term **action potential**.

The diagrams below show the sequence of events which take place along an axon as an action potential passes.

Resting potential

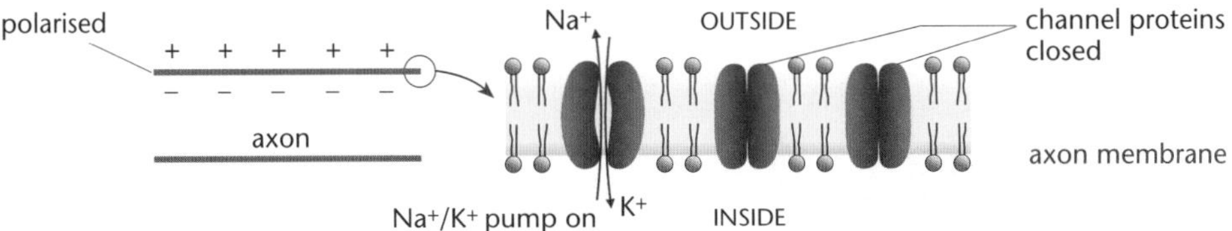

- There are 30 times more Na^+ ions on the outside of an axon during a resting potential.
- If any Na^+ ions diffuse in, then they are expelled by the '**sodium–potassium pump**'.
- The 'sodium–potassium pump' is an active transport mechanism by which a carrier protein, with ATP, expels Na^+ ions against a concentration gradient and allows K^+ ions into the axon.
- This creates a **polarisation**, i.e. there is a +ve charge on the outside of the membrane and a –ve charge on the inside.
- The potential difference is called the **resting potential** and can be measured at around –70 millivolts.

Under resting conditions, the membrane of the axon is fairly impermeable to sodium ions.

Action potential – depolarisation

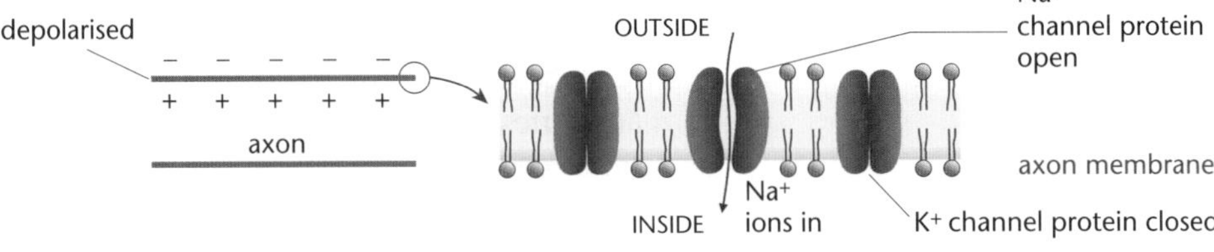

- During an action potential sodium channel proteins open to allow Na^+ ions into the axon.
- There is now a –ve charge on the outside and a +ve charge on the inside known as **depolarisation**.
- The potential difference changes to around +50 millivolts.
- The profile of the action potential, shown by an oscilloscope, is always the same.

Action potential – repolarisation

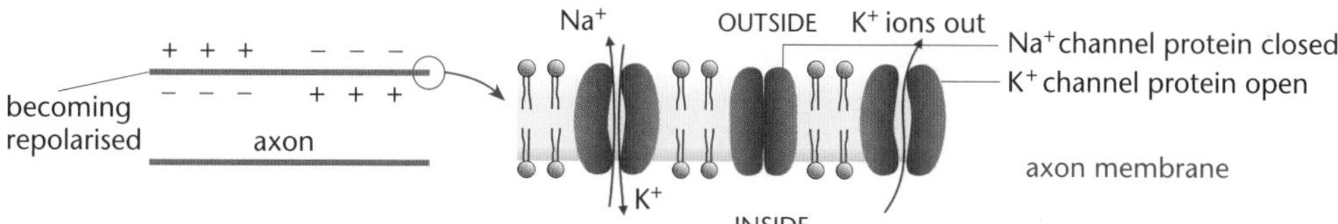

- A K^+ channel opens so K^+ ions leave the axon.
- This results in the membrane becoming polarised again.
- Any Na^+ ions that have entered during the action potential will be removed by the 'sodium–potassium pump'.

Measuring an action potential

- The speed and profile of action potentials can be measured with the help of an oscilloscope.
- The profile of the action potential for an organism always shows the same pattern, like the one shown.
- The changes in potential difference are tracked via a time base.
- Using the time base you can work out the speed at which action potentials pass along an axon as well as how long one lasts.

The diagram shows the typical profile of an action potential.

All action potentials in one neurone are the same size. A larger stimulus will increase the frequency of action potentials, not the size.

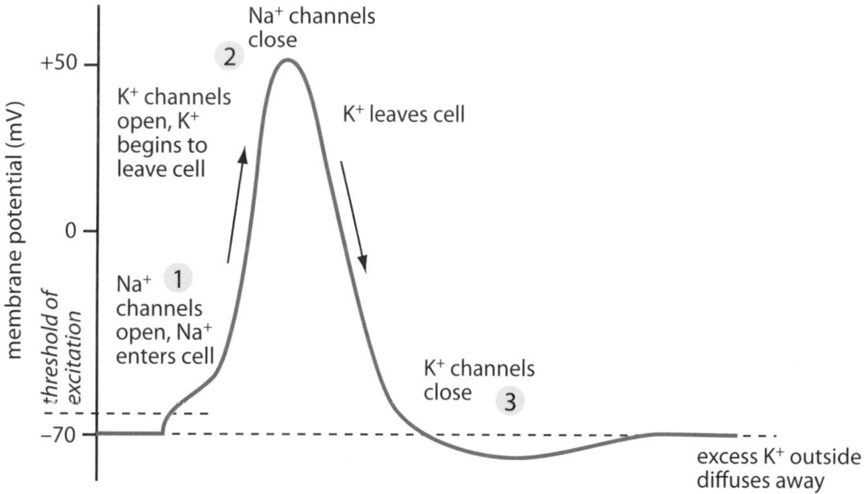

- The front of the action potential is marked by the Na^+ channels in the membrane opening.
- The potential difference increases to around +50 millivolts as the Na^+ ions stream into the axon.
- The Na^+ channels then close and K^+ channels in the membrane open.
- K^+ ions leave the axon and the membrane repolarises.
- During the **refractory period** no other action potential can pass along the axon, which makes each action potential separate or discrete.

Saltatory conduction

The reason why myelinated neurones are faster than non-myelinated neurones is that the action potential 'jumps' from one node of Ranvier to the next. This is because this is the only place where Na^+ ions can pass across the membrane. This is called **saltatory conduction**.

Progress check

What is the function of each of the following?

(a) receptor
(b) axon
(c) myelin sheath
(d) terminal dendrites

(a) Receptors respond to stimulus by producing an action potential.
(b) Transmit action potential with the help of mitochondria.
(c) Myelin sheath is a membrane enclosing fat which acts as an insulator.
(d) Terminal dendrites have motor end plates which can stimulate muscle tissue to contract.

How do neurones communicate with each other?

OCR 4.1.2

The key to links between neurones are structures known as **synapses**. Terminal dendrites branch out from neurones and terminate in **synaptic knobs**. The diagram below shows a synaptic knob separated from an interlinking neurone by a synapse.

Remember an impulse can 'cross' a synapse by chemical means and the route is in ONE direction only. They cannot go back!

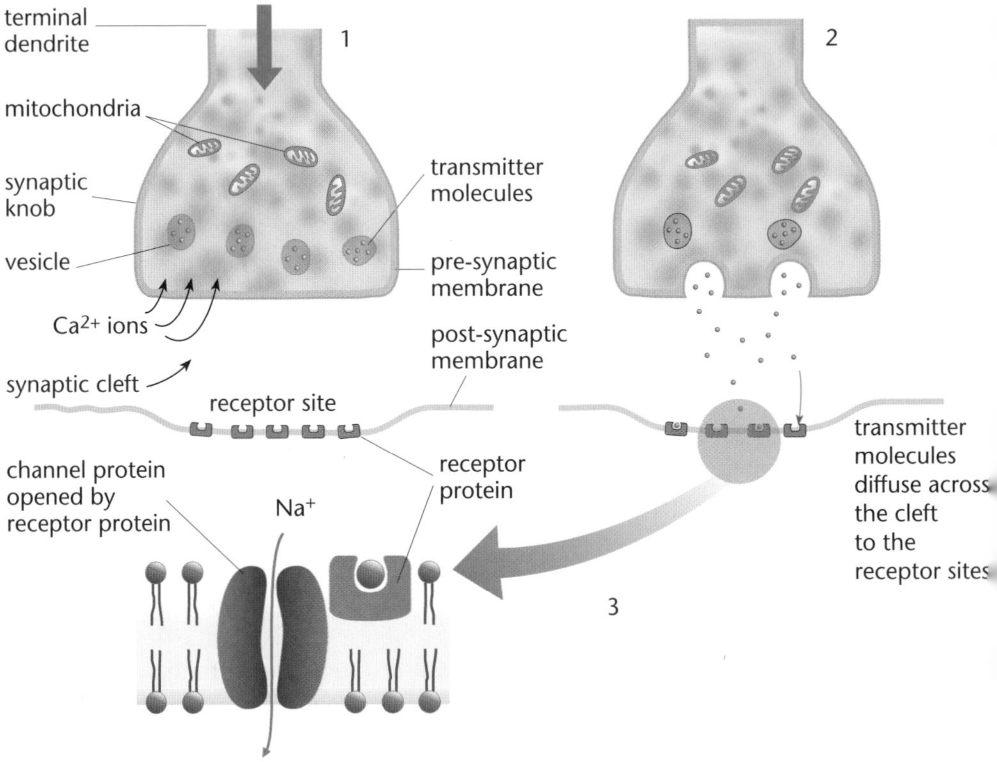

A synapse which conducts using acetylcholine is known as a cholinergic synapse.

There are **two** types of synapse:

- **excitatory** which can stimulate an action potential in a linked neurone
- **inhibitory** which can prevent an action potential being generated.

As an action potential arrives at a synaptic knob, the following sequence takes place.

- **Channel proteins** in the **pre-synaptic membrane** open to allow Ca^{2+} ions from the synaptic cleft into the synaptic knob.
- **Vesicles** then merge with the pre-synaptic membrane, so that **transmitter molecules** such as **acetylcholine** are **secreted** into the gap.
- The transmitter molecules diffuse across the cleft and bind with specific **sites** in **receptor proteins** in the **post-synaptic membrane**.
- Every receptor protein then opens a **channel protein** so that ions such as **Na^+** pass through the post-synaptic membrane into the cell.
- The Na^+ ions **depolarise** the post-synaptic membrane.
- If enough Na^+ ions enter then depolarisation reaches a **threshold level** and an **action potential** is generated in the cell.

Remember that the generation of an action potential is ALL OR NOTHING. Either enough Na^+ ions pass through the post-synaptic membrane and an action potential is generated OR not enough reach the other side, and there is no effect.

- Enzymes in the cleft then remove the transmitter substance from the binding sites, e.g. **acetylcholine esterase** removes **acetylcholine** by hydrolysing it into choline and ethanoic acid.
- Breakdown products of transmitter substances are absorbed into the synaptic knob for re-synthesis of transmitter.

Summation

A single action potential may arrive at a synaptic knob and result in some transmitter molecules being secreted into a cleft. However, there may not be enough to cause an action potential to be generated. If a series of action potentials arrive at the synaptic knob then the build up of transmitter substances may reach the threshold and the neurone will now send an action potential. We say that the neurone has 'fired' as the action potential is produced.

KEY POINT

Progress check

1 Explain the importance of summation at a synapse.

2 The diagram shows a synaptic knob.

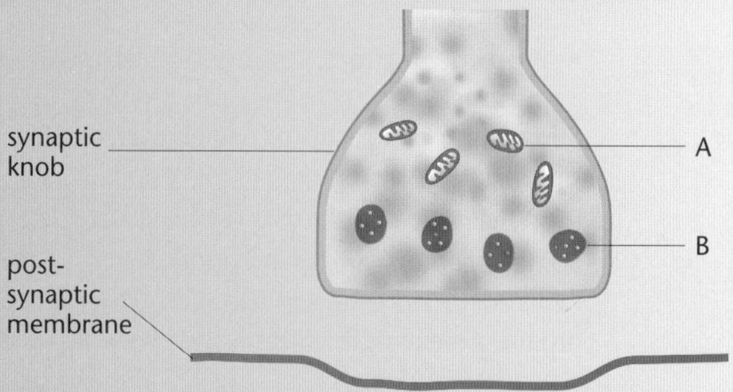

(a) Name A and B

(b) Describe the events which take place after an action potential reaches a synaptic knob and a further action potential is generated as a result.

1 A single action potential may arrive at a synaptic knob; there may not be enough transmitter molecules being secreted into a cleft to cause an action potential to be generated; a series of action potentials arrive at the synapse to build up transmitter substances to reach the threshold; the neurone will now send an action potential.

2 (a) A – mitochondria; B – vesicle

(b) Ca^{2+} ions flow into the synaptic knob; transmitter molecules such as acetylcholine are secreted into the gap; the transmitter molecules bind with sites in receptor proteins in the post-synaptic membrane; this opens channel proteins so that ions such as Na^+ pass through the post-synaptic membrane into the cell; the post-synaptic membrane is depolarised; if *enough* Na^+ ions enter a threshold level is reached and an action potential is generated in the cell.

2.3 Receptors

After studying this section you should be able to:

- *list the different types of receptors*

LEARNING SUMMARY

Types of receptors

OCR 4.1.2

The more receptors there are in a position, the more sensitive it is, e.g. the fingers have many more touch receptors than the upper arm.

Did you know?
The umbilical cord has no receptors. It can be cut without any pain.

The function of receptors is to convert the energy from different stimuli into nerve impulses in sensory neurones.

There are a range of different types of sensory cell around the body. Each type responds to different stimuli. Receptors are classified according to these different stimuli:

- **Photoreceptors**, respond to light, e.g. rods and cones in the retina.
- **Chemoreceptors**, respond to chemicals, e.g. taste buds on the tongue.
- **Thermoreceptors**, respond to temperature, e.g. skin thermoreceptors.
- **Mechanoreceptors**, respond to physical deformation, e.g. Pacinian corpuscles in the skin or hair cells in the ear.
- **Proprioreceptors**, respond to change in position in some organs, e.g. in muscles.

Stimulation of a receptor usually causes it to depolarise. This is called a **generator potential**. If this change is beyond a certain magnitude, it will trigger an action potential in a sensory neurone.

Some receptors are found individually in the body such as **Pacinian corpuscles** which detect pressure in the skin. Other receptors are gathered together into sense organs. An example of this is the eye which contains receptors called rods and cones.

2.4 Coordination by the CNS

After studying this section you should be able to:

- *outline the structure and functions of the brain and spinal cord*
- *understand the main functions of cerebrum, cerebellum, medulla oblongata and hypothalamus*

LEARNING SUMMARY

The structure and functions of the CNS

OCR 5.4.2

All neurones outside the CNS make up the peripheral nervous system.

The CNS consists of the brain and spinal cord which work together to aid the coordination of the organism. The human brain has many functions. The spinal cord takes impulses from the brain to **effectors** and in the opposite direction, impulses from **receptors** are channelled to the brain.

The brain has a complex 3D structure. The diagram below shows part of the brain structure – the major components only.

The CNS is like a motorway with impulses going in both directions.

- **Afferent neurones** take impulses **from** organs **to** the **CNS**.
- **Efferent neurones** take impulses **from** the **CNS** **to** an **organ**.

Learn these carefully. There are no marks for reversal!

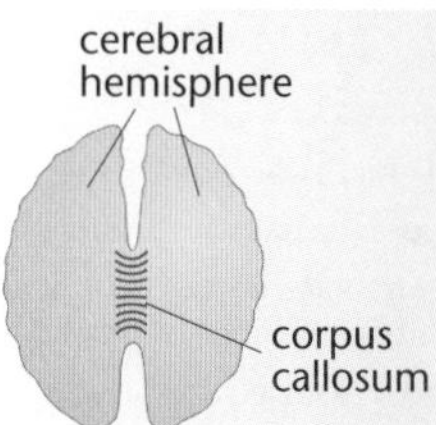

There are two cerebral hemispheres: the left and the right. Note that the right hemisphere controls the left side of the body and vice versa.

Alzheimer's disease

Neurones in the cortex of the cerebrum become progressively less able to produce neurotransmitter substances. Acetylcholine and noradrenaline are usually deficient resulting in major personality changes. The cause is often unknown, but can be genetic.

The hypothalamus is the key structure in maintaining a homeostatic balance in the body. It is like a thermostat in a house, switching the heating system on or off as internal conditions change. Similarly, it is able to control chemical levels in the blood.

The human brain

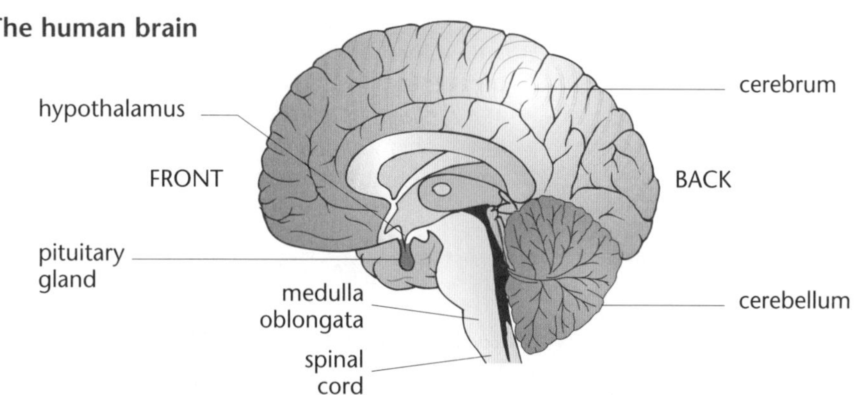

Functions of parts of the brain

Cerebrum

- **Receives sensory information** from many organs, e.g. impulses are sent from the eyes to the visual cortex at the back of the cerebrum.
- **Initiates motor activity** of many organs.
- The front of the cerebrum holds the **memory** and **intelligence** in a network of multi-polar neurones.

Cerebellum

- Has a key role in the coordination of **balance** and smooth, controlled muscular movements.
- Initiation of a movement may be by the cerebrum but the **smooth, well-coordinated** execution of the movement is only possible with the help of the cerebellum.

Medulla oblongata

- Its respiratory centre controls the rhythm of breathing with nerve connections to the intercostal muscles and the diaphragm.
- Its cardiovascular centre controls the cardiac cycle via the sympathetic and vagus nerve.
- Connects to the sino-atrial node of the heart.

Hypothalamus

- Has an exceptional blood supply.
- Many receptors are located in the blood vessel walls which supply it.
- These receptors are highly sensitive detectors which monitor:
  - temperature
  - carbon dioxide
  - ionic concentration of plasma.
- Controls body temperature by various regulatory mechanisms.
- Controls ADH secretion by the pituitary gland and is, therefore, responsible for the water content of both blood plasma and urine.

Pituitary gland

- Secretes a range of hormones and is the major control agent of the endocrine system.
- Responds to neurosecretion and release factors from the hypothalamus.
- Together with the hypothalamus, it is part of a number of negative feedback loops.
- Is the link between the nervous system and the endocrine system.

The above cover some functions of parts of the brain, but there are many more.

> The human brain consists of approximately 10^{12} neurones and all are present at birth. It is no wonder that a baby's head is proportionally large at this stage of the life cycle. During the first three months after birth, many synaptic connections are made. This is a most important developmental stage. Neurones cannot be replaced once damaged.
>
> KEY POINT

Progress check

State two functions of each of the following parts of the human brain:

(a) cerebrum (b) cerebellum (c) medulla oblongata.

(a) (i) Receives sensory impulses from the eyes to the visual cortex, enabling sight.
(ii) Controls voluntary motor activity of the leg muscles.
(b) (i) Coordinates balance, e.g. enables upright stance in humans.
(ii) Enables smooth movement, e.g. hitting a golf ball with a club straight down the fairway. (You could swing the club by voluntary control from the cerebrum but smooth coordination is by the cerebellum.)
(c) (i) Controls the rhythm of breathing with nerve connections to the intercostal muscles and the diaphragm.
(ii) Controls the heart rate via the sympathetic and vagus nerve.

2.5 Response

After studying this section you should be able to:

- *describe different types of response*
- *understand how neurones function together in a reflex arc*
- *outline the features of the autonomic nervous system*
- *describe the structure of skeletal muscle and understand the sliding filament mechanism*

LEARNING SUMMARY

Different types of response

OCR 5.4.2

In order to bring about a response, nerve impulses are sent to effectors via motor neurones. Some responses do not require conscious thought. These are called reflexes or reflex actions.

The reflex arc

How can we react quickly without even thinking about making a response?

Often the brain is not involved in the response so the time taken to respond to a stimulus is reduced. This rapid, automatic response is made possible by the **reflex arc**.

A reflex arc

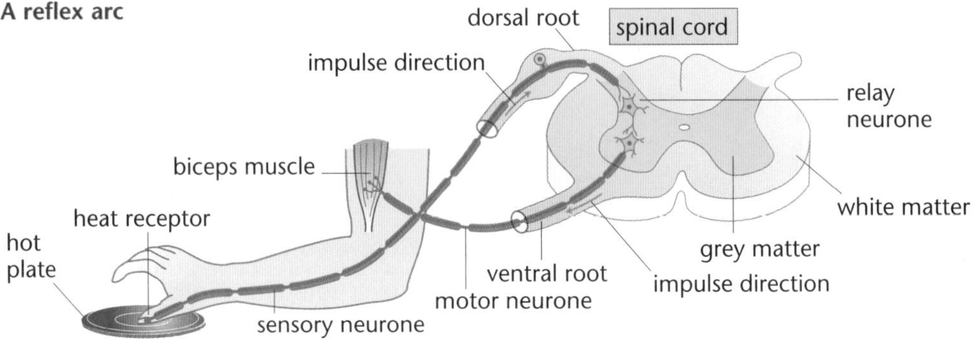

It is other afferent neurones which *finally* take impulses to the brain enabling us to be aware of the arc which has just taken place. These afferent neurones are NOT part of the reflex arc.

Features of a reflex arc

- The stimulus is detected by a **receptor**.
- As a result, an **action potential** is generated along a sensory neurone.
- The **sensory neurone** enters the spinal cord via the **dorsal root** and **synapses** onto a **relay neurone** / intermediate neurone.
- This intermediate neurone synapses onto a **motor neurone** which in turn conducts the impulse to **a muscle** via its **motor end plates**.
- The muscle contracts and the arm instantly **withdraws** from the stimulus before any harm is done.
- The complete list of events takes place so quickly because the impulses do not, initially, go to the brain! The complete pathway to the muscle conducts the impulse so rapidly, before the brain receives any sensory information.

> Reflexes have a high survival value because the organism is able to respond so rapidly. Additionally, they are always automatic. There are a range of different reflexes, e.g. iris/pupil reflex and saliva production.
>
> **KEY POINT**

The iris/pupil reflex

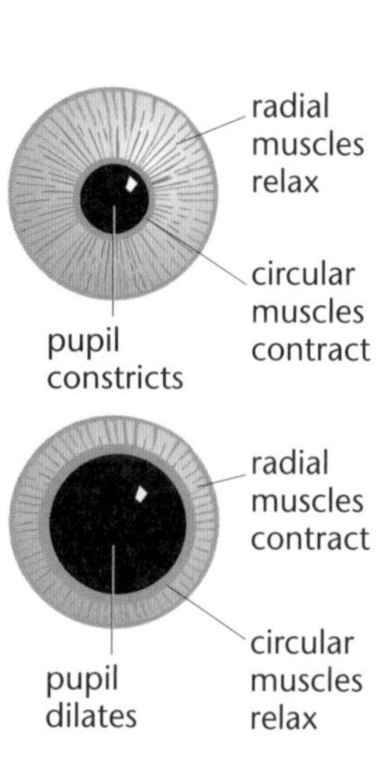

This response does involve the brain, but because conscious thought is not involved, it is still classed as a reflex.

The diagrams in the margin show the two extremes of pupil size.

- The amount of light entering the eye is detected by **receptors** in the **retina**.
- Reflex pathways lead to the **circular** and **radial** muscles of the **iris**.
- **High-intensity** light activates the **circular muscles** of the iris to **contract**; as the radial muscles relax so the **pupil gets smaller**. (The advantage of this is too much light does not enter and so does not damage the retina.)
- **Low-intensity** light activates the **radial muscles** of the iris to **contract**; as the circular muscles relax so the **pupil gets wider**. (The advantage of this is that the eye allows enough light to see.)
- A balance between the two extremes is achieved across a gradation of light conditions.

Autonomic nervous system

Many reflex actions are controlled by the autonomic nervous system. This is the part of the nervous system which controls our involuntary activities, e.g. the control of the heart rate. It is divided into two parts, the **sympathetic system** and the **parasympathetic system**. Each system has a major nerve from which smaller nerves branch out into key organs. In some ways, the two systems are **antagonistic** to each other but in other ways, they have specific functions not counteracted by the other. The table below shows all of the main facts for each system.

This table of features shows some key points for the autonomic system. ALERT! They are difficult to learn because of the lack of logic in the 'pattern' of functions. Take time to revise this properly because many candidates mix up the features of one system with another.

	Autonomic nervous system	
Feature	*Sympathetic*	*Parasympathetic*
Nerve (example)	sympathetic nerve	vagus nerve
Transmitter substance at synapses	noradrenaline	acetylcholine
Heart rate	speeds up	slows down
Iris control	dilates	constricts
Saliva	————	flow stimulated
Gut movements	slows down	speeds up
Sweating	sweat production stimulated	————
Erector pili muscles	contracts erector pili muscles	————

> Remember that all of the above functions take place without thought. The system is truly involuntary.
>
> KEY POINT

Muscles as effectors

OCR 5.4.2

The body has a number of different effectors, but for most responses the effector is a muscle. There are three types of muscle in the body:

- skeletal/striated or voluntary muscle
- visceral/smooth or involuntary muscle
- cardiac muscle.

Smooth muscle is controlled by the autonomic nervous system, but skeletal muscle is controlled by motor neurones of the somatic nervous system.

How do motor neurones control muscle tissue?

The link to muscle tissue is by **motor end plates** which have close proximity to the sarcoplasm of the muscle tissue. The motor end plates have a greater surface area than a synaptic knob, but their action is very similar to the synaptic transmission described on page 36. Action potentials result in muscle contraction.

No contraction would take place without the acetylcholine transmitter being released from the motor end plate. When the sarcolemma (membrane) reaches the threshold level, then the action potential is conducted throughout the sarcolemma. Contraction is initiated!

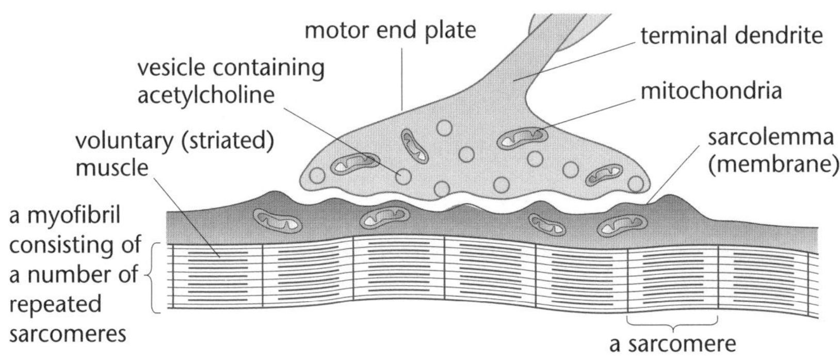

Skeletal muscle is also known as striated or striped muscle. The structure of a single muscle unit, the sarcomere, shows the striped nature of the muscle.

The sarcomere

A sarcomere showing bands

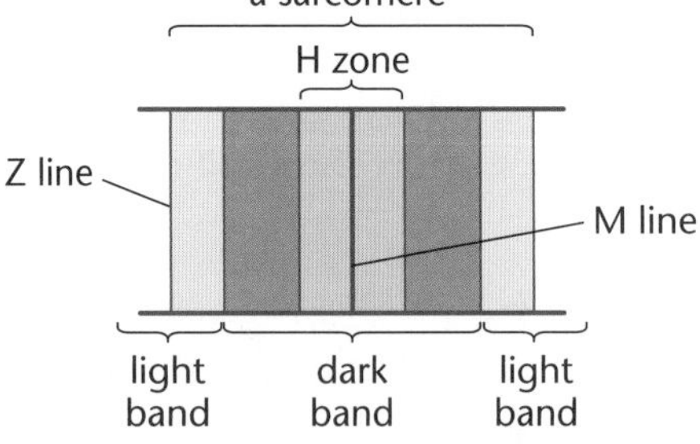

A sarcomere showing filaments

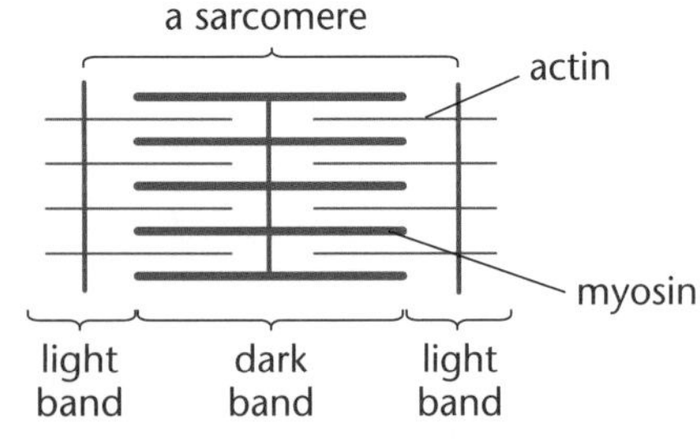

- The sarcomere consists of different filaments, thin ones (**actin**) and thick ones (**myosin**).
- These filaments form bands of different shades:
  - light band (I bands) – just actin filaments
  - dark band (A bands) – just myosin filaments or myosin plus actin.
- During contraction, the filaments slide together to form a shorter sarcomere.
- As this pattern of contraction is repeated through 1000s of sarcomeres, the whole muscle contracts.

- Actin and myosin filaments slide together because of the formation of cross bridges which alternatively build and break during contraction.
- Cross bridge formation is known as the 'ratchet mechanism'.

How does the 'ratchet mechanism' work?

To answer this question, the properties of actin and myosin need to be considered. The diagram below represents an actin filament next to a myosin filament. Many 'bulbous heads' are located along the myosin filaments (just one is shown). Each points towards an actin filament.

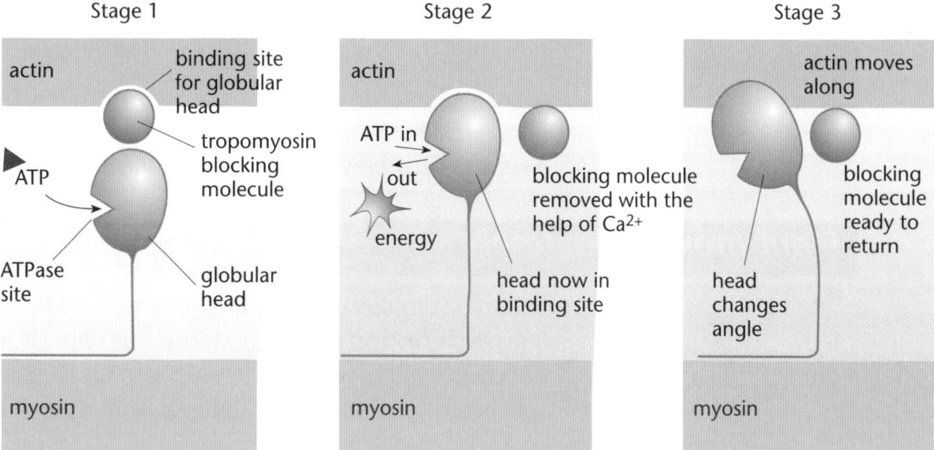

The sequence of the ratchet mechanism

- Once an action potential is generated in the muscle tissue then **Ca^{2+} ions** are released from the **reticulum**, a structure in the **sarcoplasm**.
- Part of the **globular head** of the myosin has an **ATPase** (enzyme) site.
- Ca^{2+} ions activate the myosin head so that this ATPase site hydrolyses an ATP molecule, **releasing energy**.
- Ca^{2+} ions also bind to **troponin** in the actin filaments, this in turn removes **blocking molecules** (**tropomyosin**) from the actin filament.
- This exposes the **binding sites** on the **actin** filaments.
- The globular heads of the myosin then bind to the newly exposed sites forming **actin–myosin cross bridges**.
- At the stage of energy release the myosin **heads change angle**.
- This change of angle moves the actin filaments towards the centre of each sarcomere and is termed the **power stroke**.
- More ATP binds to the myosin head, effectively causing the cross bridge to become straight and the tropomyosin molecules once again block the actin binding sites.
- The myosin is now 'cocked' and ready to repeat the above process.
- Repeated cross bridge formation and breakage results in a rowing action shortening the sarcomere as the filaments slide past each other.

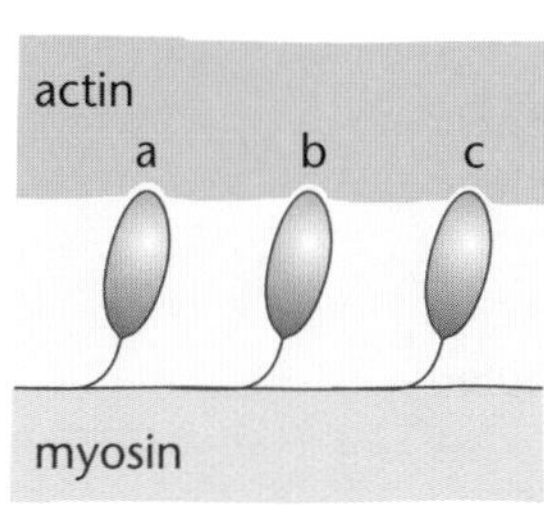

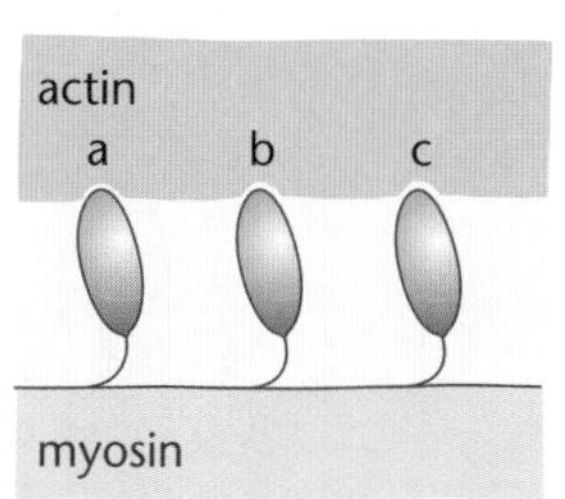

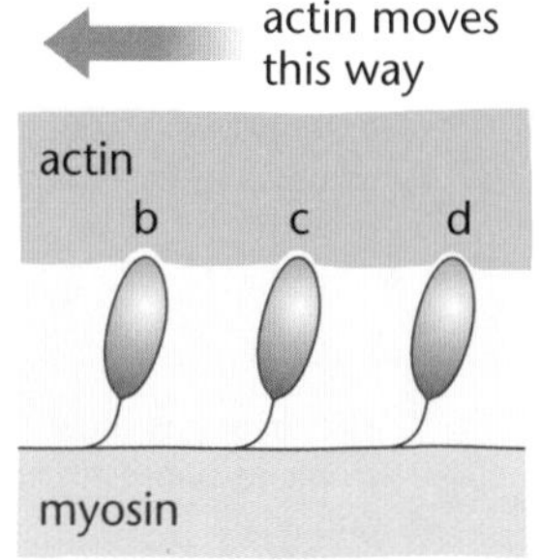

How does skeletal muscle produce movement?

Motor neurones control the skeletal muscle via motor end plates. The skeletal muscles move the bones via their tendon attachments. The muscles work in **antagonistic** pairs, i.e. **opposing** each other. In the arm, when the biceps contracts the forearm is lifted. At the same time the triceps relaxes. If the forearm is to be lowered then the triceps contracts and the biceps relaxes.

2.6 Plant sensitivity

After studying this section you should be able to:

- *understand the range of tropisms which affect plant growth*
- *understand how auxins, gibberellins and cytokinins control plant growth*
- *understand how phytochromes control the onset of flowering in plants*

LEARNING SUMMARY

Plant growth regulators

OCR 5.4.1

External stimuli such as light can affect the direction of plant growth. A **tropism** is a **growth response** to an external stimulus. It is important that a plant grows in a direction which will enable it to obtain maximum supplies. Plant regulators are substances produced in minute quantities and tend to interact in their effects.

Growth response to light is **phototropism**
Growth response to gravity is **geotropism**
Growth response to water is **hydrotropism**
Growth response to contact is **thigmotropism**

Tropisms can be positive (towards) or negative (away from).

Phototropism

This response is dependent upon the stimulus – light affecting the growth regulator, **auxin** (indoleacetic acid).

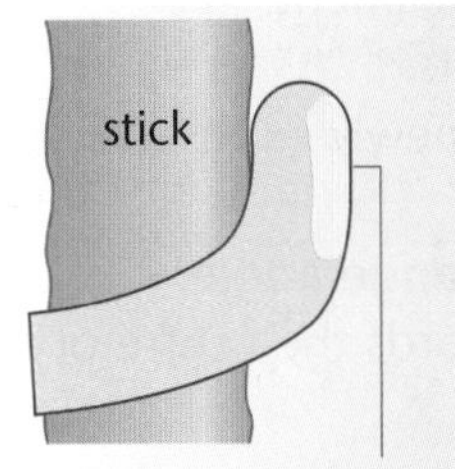

Thigmotropism helps a climbing plant like the runner bean to grow in a twisting pattern around a stick. Auxin is redistributed away from the contact point so the outer cells elongate giving a stronger outer growth.

Auxin and growing shoots

Auxin is produced by cells undergoing mitosis, e.g. growing tips. If a plant shoot is illuminated from one side then the auxin is redistributed to the side furthest from the light. This side grows more strongly, owing to the elongation of the cells, resulting in a bend towards the light. The plant benefits from increased light for photosynthesis. Up to a certain concentration, the degree of bending is proportional to auxin concentration.

The diagrams show tropic responses to light and auxin.

Tropisms in response to light from different directions **Tropism in response to auxin**

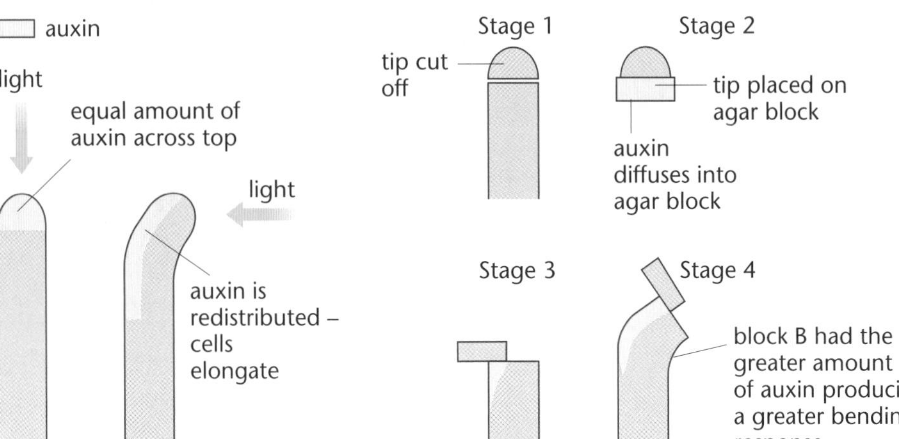

Auxin research

Many investigations of auxins have taken place using the growing tips of grasses. Where a growing tip is removed and placed on an agar block, auxin will diffuse into the agar. Returning the block to the mitotic area stimulates increased cell elongation to the cells receiving a greater supply of auxin.

Is the concentration of auxin important?

It is important to consider the implications of the concentration of auxin in a tissue. The graph below shows that at different concentrations, auxin affects the shoot and the root in different ways.

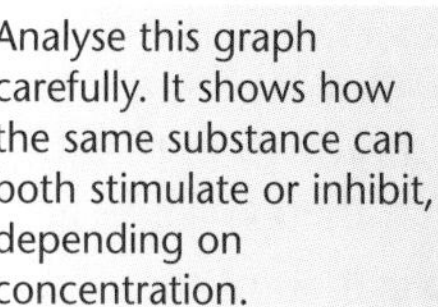

Analyse this graph carefully. It shows how the same substance can both stimulate or inhibit, depending on concentration.

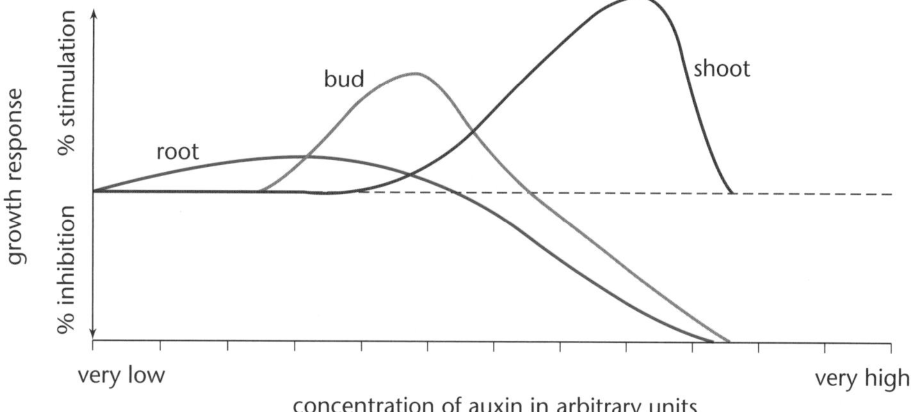

The graph shows:

- auxin has no effect on a shoot at very low concentration
- at these very low concentrations root cell elongation is stimulated
- at higher concentrations the elongation of shoot cells is stimulated
- at these higher concentrations auxin inhibits the elongation of root cells.

Auxin and root growth

The graph above shows that auxin affects root cells in a different way at different concentrations. At the root tip, auxin accumulates at a lower point because of gravity. This inhibits the lower cells from elongating. However, the higher cells at the tip have a low concentration of auxin and do elongate. The net effect is for the stronger upper cell growth to bend the root downwards. The plant therefore has more chance of obtaining more water and mineral ions.

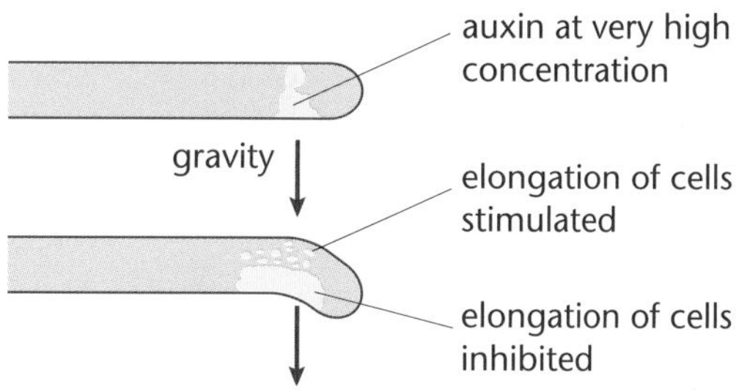

Plant growth regulators

In your examination, look out for data which will be supplied, e.g. the growth regulator gibberellin may be linked to falling starch levels in a seed endosperm and increase in maltose. Gibberellin has stimulated the enzymic activity.

Hormone	*Some key functions*
auxin	increased cell elongation, suppression of lateral bud development
gibberellin	cell elongation, ends dormancy in buds, promotes germination of seeds by activating hydrolytic enzymes such as amylase (food stores are mobilised!)
cytokinin	increased cell division, increased cell enlargement in leaves inhibits leaf fall
ethene	promotes ripening of food, stimulates leaf fall

Commercial uses of plant growth regulators

Now that many functions of plant growth regulators have been discovered, gardeners and farmers can use them to manipulate the growth of plants.

Growth regulator	*Commercial use*	*Details*
Auxins	Taking cuttings	Cuttings are taken from shoots and dipped into rooting powder containing auxins. This stimulates the growth of lateral roots.
	Seedless fruits	If the flowers are treated with auxins, the fruit may develop without fertilisation. It will therefore not contain a seed.
	Weedkillers	Synthetic auxins can be used as selective herbicides (weedkillers). They stimulate rapid growth in the shoots, which then collapse and die.
Gibberillins	Delaying fruit drop	Application of the gibberillins can delay fruit drop, making the fruits riper and preventing damage.
Cytokinins	Tissue culture	Promotes bud and shoot growth in small pieces of plant tissue.
Ethene	Promotes fruit ripening	Fruits can be picked and transported unripe to prevent damage and then ripened with ethene.

Sample question and model answer

(a) The diagram shows a neurone.

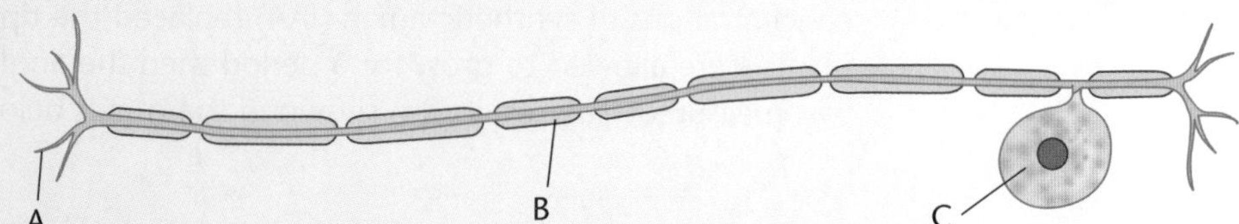

(i) What type of neurone is shown in the diagram? [1]

sensory neurone

(ii) What structure would be found at A? [1]

a receptor

(iii) Where precisely in the body is structure C found? [1]

in the dorsal root ganglion

(iv) Structure B is covered by a myelin sheath.
Explain the function of the myelin sheath. [3]

The myelin sheath insulates the neurone;
Ions can only pass into the neurone at the nodes of Ranvier;
This speeds up the transmission of the nerve impulse.

(b) The diagram shows a synapse.

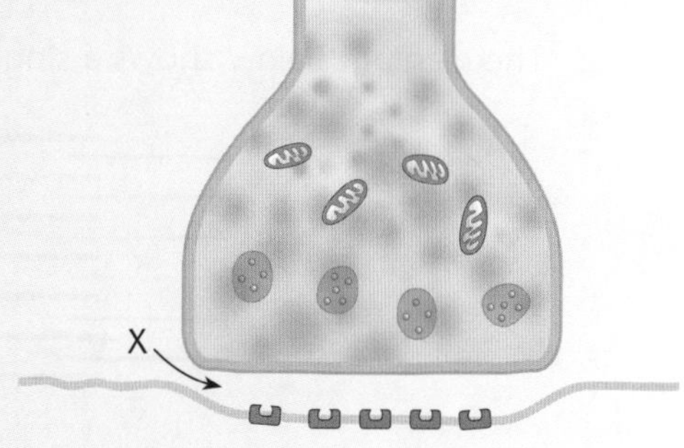

(i) What is contained inside the synaptic vesicles? [1]

Neurotransmitter / Acetylcholine / transmitter molecules

(ii) What is the name of the gap marked X? [1]

Synaptic cleft

(iii) Why are there so many mitochondria in the end of the neurone? [2]

To produce ATP;
For the production of neurotransmitter;
To reabsorb the neurotransmitter;
For exocytosis

Practice examination questions

1 The growing tips were removed from oat stems. Agar blocks containing different concentrations of synthetic auxin (IAA) replaced the tips on the oat stems. The plants were allowed to grow for a period then the angle of curvature of the stems was measured. The results are shown in the graph below.

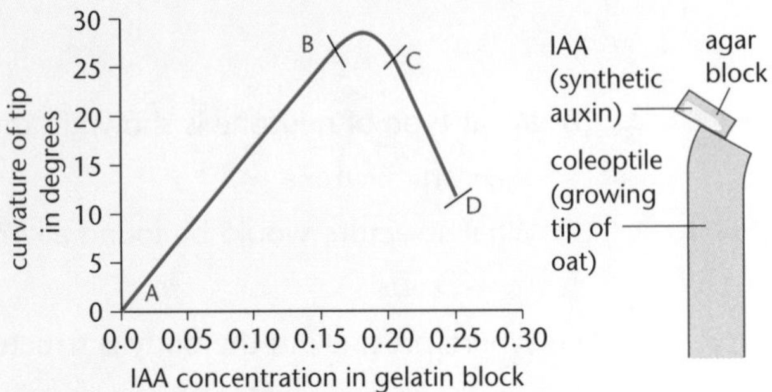

(a) What is the relationship between IAA concentration and curvature of the stem between points:

(i) A and B [1]

(ii) C and D? [1]

(b) Explain how IAA causes a curvature in the oat stems. [2]

(c) Explain the effect a much higher concentration of IAA would have on the curvature of oat stems. [2]

[Total: 6]

2 The diagram below shows a single sarcomere just before contraction.

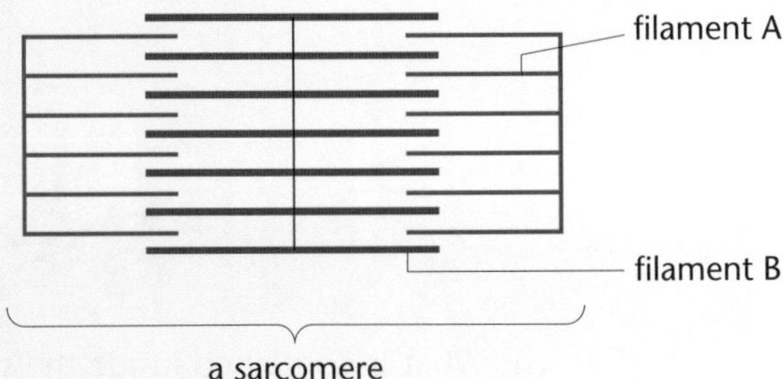

(a) Name filaments A and B. [2]

(b) What stimulus causes the immediate contraction of a sarcomere? [1]

(c) What happens to each type of filament during contraction? [2]

[Total: 5]

3 The diagram below shows the profile of an action potential.

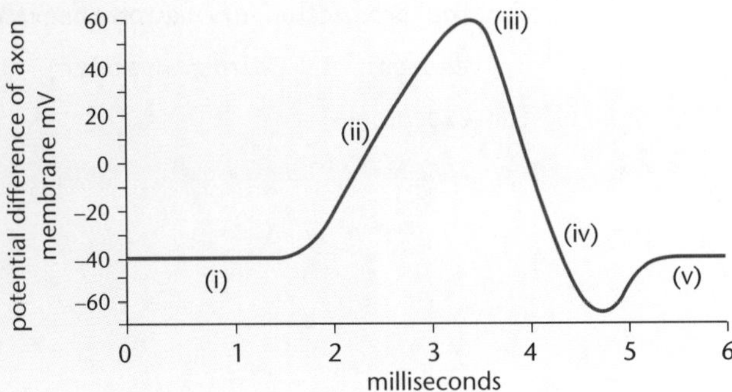

Explain what happens in the axon at each stage shown on the diagram. [10]

[Total: 10]

Chapter 3

Homeostasis

The following topics are covered in this chapter:

- Hormones
- Regulation of blood sugar level
- Temperature control in a mammal
- The kidneys

3.1 Hormones

LEARNING SUMMARY

After studying this section you should be able to:

- define homeostasis
- describe the route of hormones from source to target organ
- understand that hormones and nerves contribute to homeostasis

The endocrine system

OCR 4.1.1/3

The endocrine system secretes a number of chemicals known as **hormones**. Each hormone is a substance produced by an **endocrine gland**, e.g. adrenal glands produce the hormone adrenaline. Each hormone is **transported in the blood** and has a **target organ**. Once the target organ is reached, the hormone **triggers a response** in the organ. Many hormones do this by **activating enzymes**. Others **activate genes**, e.g. steroids.

The great advantage of homeostasis is that the conditions in the environment fluctuate but conditions in the organism remain stable.

KEY POINT

The endocrine and nervous systems both contribute to **coordination** in animals. They help to regulate internal processes. **Homeostasis** is the maintenance of a **constant internal environment**. Nerves and hormones have key roles in the maintenance of this **steady internal state.** Levels of pH, blood glucose, oxygen, carbon dioxide and temperature all need to be controlled.

Parts of the human endocrine system (both male and female organs shown!)

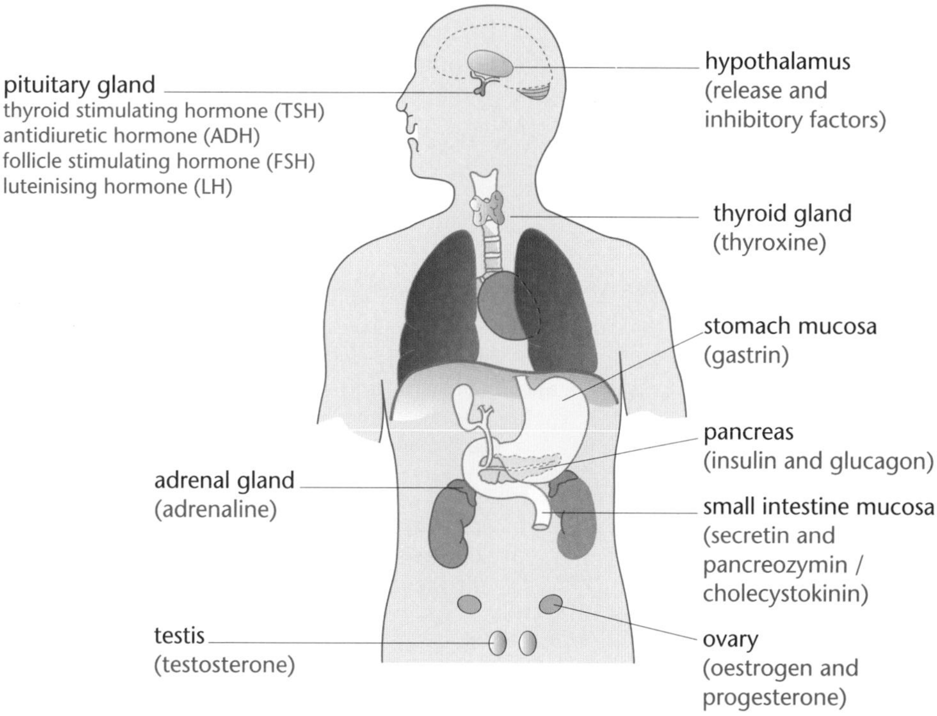

The production of adrenaline prepares the body for 'fight or flight'. It increases the heart rate and breathing rate. It also converts glycogen into glucose.

How does a hormone trigger a cell in a target organ?

When a cell responds to a hormone or a nerve impulse or a chemical, it is called **cell signalling**.

Hormones are much slower in eliciting a response than the nervous system. Rather than having an effect in milliseconds like nerves, hormones take longer. However, effects in response to hormones are often **long lasting**.

The diagram below shows one mechanism by which hormones activate target cells.

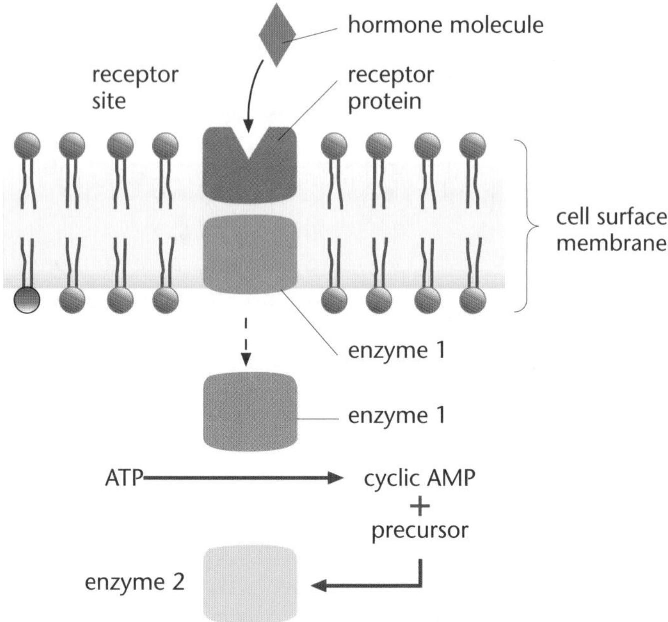

Look carefully at this mechanism! Just **ONE hormone molecule** arriving at the cell releases an enzyme which can be used **many** times. In turn, another enzyme is produced which can be used **many** times. One hormone molecule leads to **amplification**. This is a cascade effect.

Hormones that are proteins work in this way because they are unable to enter the cell. Steroid hormones, (e.g. oestrogen) can pass through the cell membrane as they are lipid soluble.

Nerves and hormones working together

The nervous system and the endocrine system work together in the body to achieve homeostasis. An example of this cooperation is in the control of the heart rate. Cardiac muscle in the heart is myogenic. This means that it will contract without a nerve impulse. The rate is set by the pacemaker in the SA node but this can be adjusted by nerves and hormones. This allows the body to match the output of the heart to the demands of exercise.

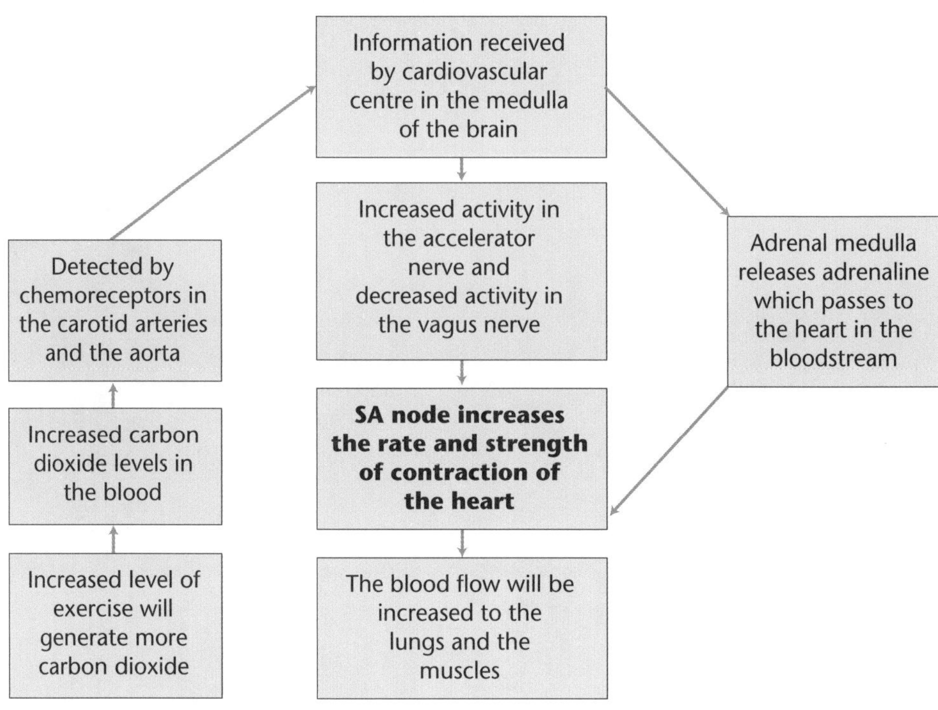

3.2 Temperature control in a mammal

After studying this section you should be able to:

LEARNING SUMMARY

- *outline the processes which contribute to temperature regulation in a mammal*
- *understand how nervous and endocrine systems work together to regulate body temperature*
- *understand how internal processes are regulated by negative feedback*

What are the advantages of controlling body temperature?

OCR 4.1.1

It is advantageous to maintain a constant body temperature so that the enzymes which drive the processes of life can function at an optimum level.

- **Endothermic** (warm blooded) animals can maintain their core temperature at an optimal level. This allows internal processes to be consistent. The level of activity of an endotherm is likely to fluctuate less than an ectotherm.
- **Ectothermic** (cold blooded) animals have a body temperature which fluctuates with the environmental temperature. As a result there are times when an animal may be vulnerable due to the enzyme-driven reactions being slow. When a crocodile (ectotherm) is in cold conditions, its speed of attack would be slow. When in warm conditions, the attack would be rapid.

Once the blood temperature decreases, the heat gain centre of the hypothalamus is stimulated. This leads to a rise in blood temperature which, in turn, results in the heat loss centre being stimulated. This is negative feedback! The combination of the two, in both directions, contributes to homeostasis.

How is temperature controlled in a mammal?

The key structure in homeostatic control of all body processes is the **hypothalamus**. The regulation of temperature involves thermoreceptors in the skin, body core and blood vessels supplying the brain, which link to the hypothalamus.

The **hypothalamus** has **many functions**. It controls thirst, hunger, sleep and it stimulates the production of many hormones other than those required for temperature regulation.

The diagram below shows how the peripheral nerves, hypothalamus and pituitary gland integrate nervous and endocrine glands to regulate temperature.

Temperature regulation model

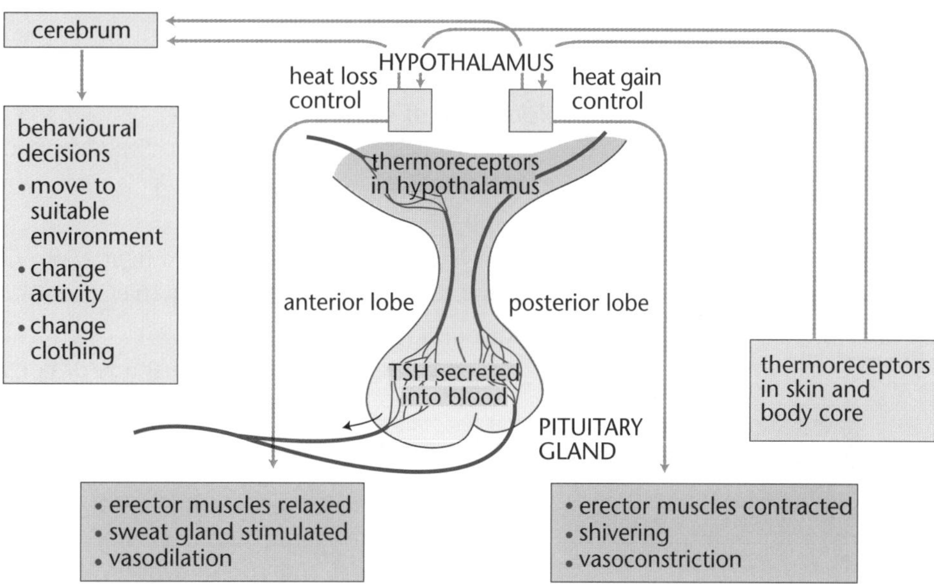

If there is an increase in core temperature then the hypothalamus stimulates greater heat loss by:

- vasodilation (dilation of the skin arterioles)
- relaxing of erector-pili muscles so that hairs lie flat
- more sweating
- behavioural response in humans to change to thinner clothing.

When the hypothalamus receives sensory information **heat loss** or **heat gain** control results.

A capillary bed

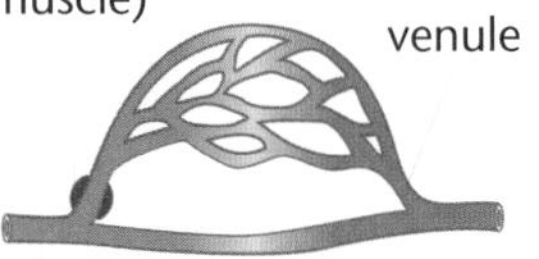

A **fall** in temperature results in the following control responses.

Vasoconstriction

- Arteriole control is initiated by the hypothalamus which results in efferent neurones stimulating constriction of the arteriole sphincters of skin capillary beds.
- This deviates blood to the core of the body, so less heat energy is lost from the skin.

Contraction of the erector-pili muscles

- Erector-pili muscle contraction is initiated in the hypothalamus being controlled via efferent neurones.
- Hairs on skin stand on end and trap an insulating layer of air, so less heat energy is lost from the skin.

Sweat reduction

- The sweat glands control is also initiated in the hypothalamus, and is controlled via efferent neurones.
- Less heat energy is lost from the skin by evaporation of sweat.

Shivering

- Increased muscular contraction is accompanied by heat energy release.

Behavioural response

- A link from the hypothalamus to the cerebrum elicits this voluntary response.
- This could be to switch on the heating, put on warmer clothes, etc.

Increased metabolic rate

- The hypothalamus produces a release factor substance.
- This stimulates the anterior part of the pituitary gland to secrete TSH (thyroid stimulating hormone).
- TSH reaches the thyroid via the blood.
- Thyroid gland is stimulated to secrete thyroxine.
- Thyroxine increases respiration in the tissues increasing the body temperature.

Once a higher thyroxine level is detected in the blood the release factor in hypothalamus is inhibited so TSH release by the pituitary gland is prevented. This is **negative feedback.**

Note
(a) the outline for heat loss methods does not show the nerve connections. Efferent neurones are again coordinated via the hypothalamus!
(b) heat is lost from the skin via a combination of **conduction, convection** and **radiation.**

An **increase** in body temperature results in almost the **opposite** of each response described for a fall of temperature.

Vasodilation

- Arterioles of capillary beds dilate allowing more blood to skin capillary beds.

Relaxation of erector-pili muscles

- Hairs lie flat, no insulating layer of air trapped.
- More heat loss of skin.

Sweat increase

- More sweat is excreted so more heat energy from the body is needed to evaporate the sweat, so the organism cools down.

Behavioural response

- This could be to move into the shade or consume a cold drink.

Progress check

Hormone X stimulates the production of a substance in a cell of a target organ. The following statements outline events which result in the production of the substance but are in the wrong order. Write the correct order of letters.

A Hormone X is transported in the blood.
B Hormone X binds with a receptor protein in the cell surface membrane.
C The enzyme catalyses a reaction, forming a product.
D Hormone X is secreted by a gland.
E This releases an enzyme from the cell surface membrane.

D, A, B, E, C.

3.3 Regulation of blood sugar level

After studying this section you should be able to:

- *understand the control of blood glucose levels in a person*
- *describe the sites of insulin and glucagon secretion*
- *explain the functions of insulin and glucagon*

LEARNING SUMMARY

Why is it necessary to control the amount of glucose in the blood?

OCR 4.2.1, 4.1.3

Glucose molecules are needed to supply energy for every living cell. The level in the blood must be high enough to meet this need (90 mg per 100 cm^3 blood). This level needs to be maintained at a constant level, even though a person may or may not have eaten. High levels of glucose in the blood would cause great problems. Hypertonic blood plasma would result in water leaving the tissues by osmosis. Dehydration of organs would result in a number of symptoms.

Blood glucose regulation

Remember that the pancreas also produces enzymes. The hormones are released from the islets of Langerhans, which are isolated groups of cells.

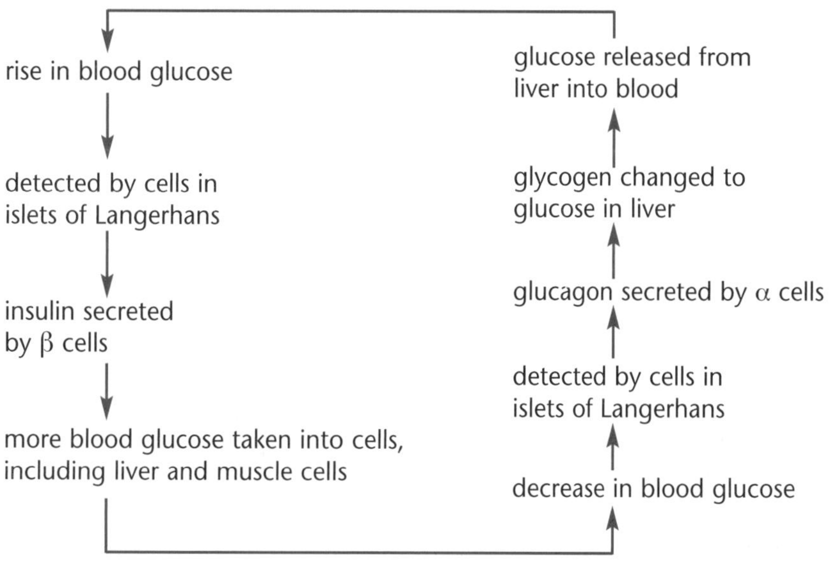

Positive feedback is rare in the body. It means that any change in a variable brings about a response that changes the variable even more.

Negative feedback
Blood glucose regulation is an example of negative feedback. Any change in glucose level initiates changes which will result in the return of the original level – **balance is achieved.**

KEY POINT

Never state that insulin changes glucose to glycogen. It allows glucose into the liver where the enzyme glycogen synthase catalyses the conversion.

Insulin

- Is secreted into the blood due to stimulation of pancreatic cells by a **high concentration** of glucose in the blood.
- Is produced by the β **cells** of the **islets of Langerhans** in the pancreas.
- Binds to receptor proteins in cell surface membranes activating carrier proteins to **allow glucose entry** to cells.
- Allows **excess glucose** molecules into the liver and muscles where they are converted into **glycogen** (a storage product), and some fat.

Glucagon

- Is secreted into the blood due to stimulation of pancreatic cells by a **low concentration** of glucose in the blood.
- Is produced by the α **cells** of the **islets of Langerhans** in the pancreas.
- Stimulates the conversion of glycogen to **glucose**.

Diabetes

There are two types of this condition.

Type 1

- The pancreas fails to produce enough insulin.
- After a meal when blood glucose level increases dramatically, the level remains high.
- High blood glucose causes hyperglycaemia.
- Kidneys, even though they are healthy, cannot reabsorb the glucose, resulting in glucose being in the urine.
- Symptoms include dehydration, loss of weight and lethargy.

What is the answer?

- Insulin injections and a carbohydrate controlled diet.

Type 2

- This form of diabetes usually occurs in later life.
- Insulin is still produced but the receptor proteins on the cell surface membranes may not work correctly.
- Glucose uptake by the cells is erratic.
- Symptoms are similar to those for type 1 but are mild in comparison.

What is the answer?

- Dietary control including low carbohydrate intake.

More liver functions

The role of the liver in its production of bile, as well as the storage and break down of glycogen has been highlighted. The liver does so much more!

Children need 10 essential amino acids (adults need 8). From these they can make different ones by transamination in the liver.

Transamination

This is the way an R group of a keto acid is transferred to an amino acid. It replaces the existing R group with another and a new amino acid is formed.

CH_3 amino acid + C_2H_5 keto acid → C_2H_5 amino acid + CH_3 keto acid

Note the changes in the 'R' group of each acid.

The liver has many functions including:

- detoxification of poisonous substances
- heat production.

Deamination

This process is necessary to lower the level of **excess amino acids**. They are produced when proteins are digested. Nitrogenous materials have a high degree of toxicity, so the level in the blood must be limited.

$$2NH_2-\underset{H}{\overset{R}{C}}-COOH + O_2 \rightarrow 2\underset{O}{\overset{R}{C}}-COOH + 2NH_3$$

amino acid + oxygen → keto acid + ammonia

Both deamination and the ornithine cycle are needed to process excess amino acids. Remember that urea is a less toxic substance. The **liver makes** it but the **kidneys** help to **excrete** it.

Ornithine cycle

Ammonia is immediately taken up by ornithine to help make a less toxic substance, urea.

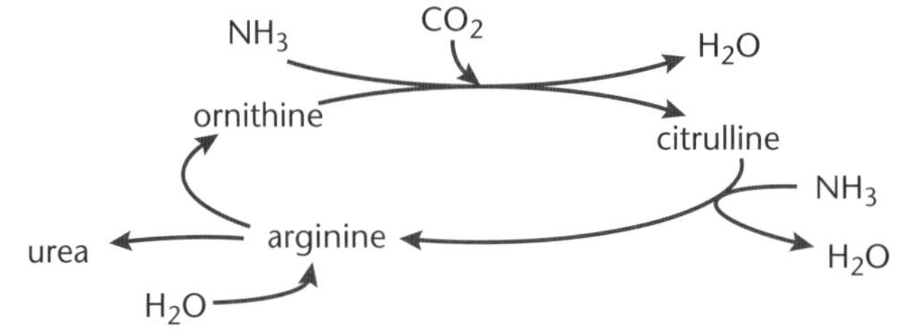

3.4 The kidneys

After studying this section you should be able to:

- *describe the structure and functions of a nephron*
- *understand the processes of ultrafiltration and reabsorption*
- *understand the countercurrent multiplier*

LEARNING SUMMARY

Kidney structure and function

Remember that excretion is the removal from the body of waste products produced by metabolism.

Each kidney has three major regions: the **cortex**, **medulla** and **pelvis**. The renal artery takes blood into a kidney where it is filtered to remove potentially toxic material. Useful substances leave the blood as well as toxic ones but are reabsorbed back into the blood. Toxic substances such as urea leave the kidneys and enter the bladder, via the ureters.

The diagram shows one nephron of the many thousands of nephrons in each kidney.

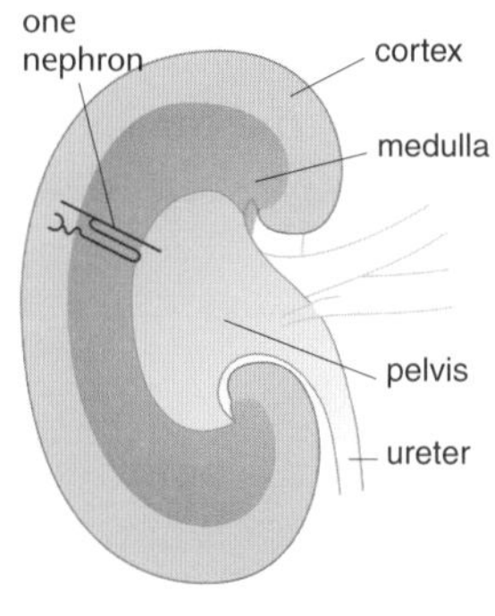

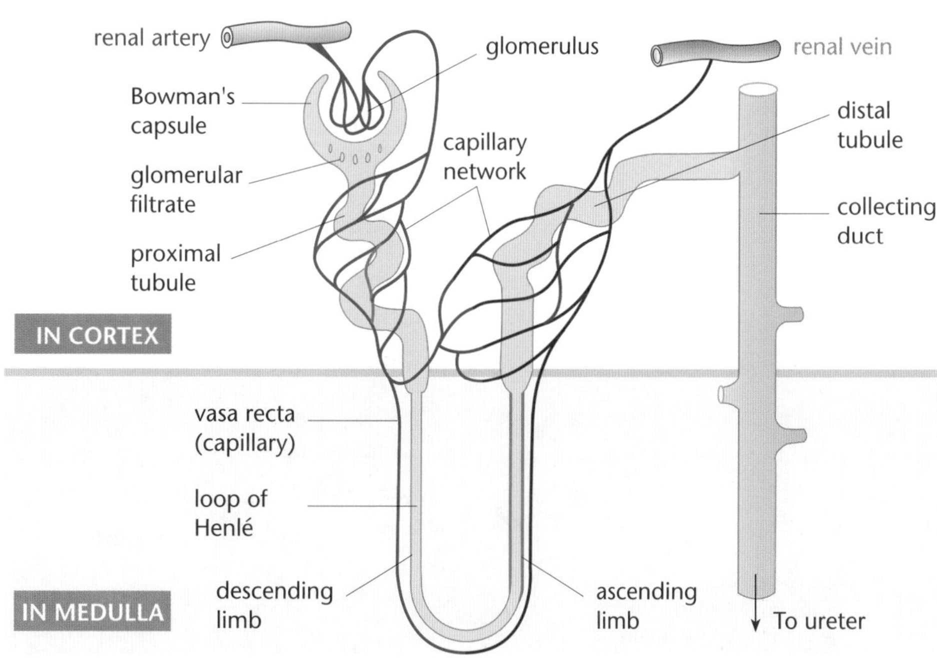

How does a nephron function?

- Blood arrives at the **glomerulus** from the renal artery.
- The **blood pressure is very high** as a result of:
  - contraction of the left ventricle of the heart
  - the arteriole leading to glomerular capillaries is wider than the venule leaving them
  - high resistance of the interface between glomerular capillaries and the inner wall of the renal (Bowman's) capsule.
- **Glomerular filtrate** is forced into the nephron; this is known as **ultrafiltration**.
- Glomerular filtrate includes **urea**, **glucose**, **water**, **amino acids** and **mineral ions**.
- Selective reabsorption takes place in the **proximal tubule** resulting in substances such as glucose being returned to the blood.
- 100% of glucose and 80% of water are reabsorbed at the proximal tubule.
- Urea continues through the tubule to the collecting duct and finally down a ureter to be excreted from the bladder.
- Further reabsorption of substances can take place at the distal tubule.

The selective property of the renal membrane.

What do not leave the blood because they are too large?

Most proteins, red and white blood cells.

Ultrafiltration

Also known as pressure filtration, ultrafiltration relies on the properties of the capillaries and the inner wall of the renal (Bowman's) capsule.

In your examination, you may be requested to label a diagram. Test yourself!

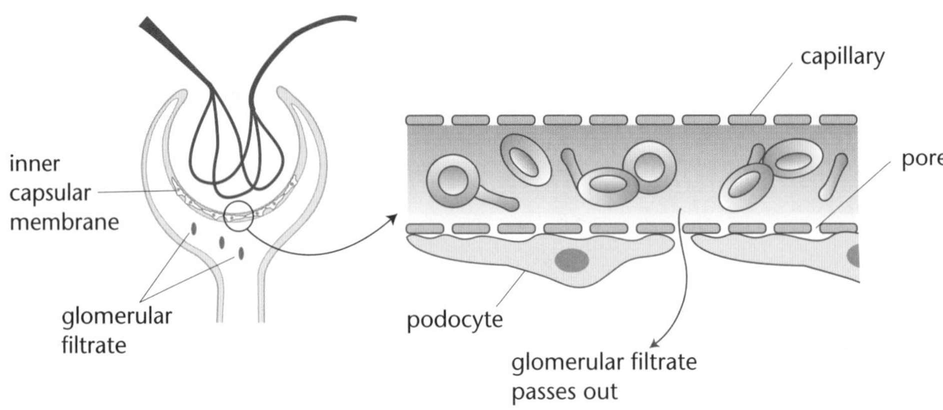

Learn the explanations in the bullet points. Bullet points in this book may resemble your examiner's mark scheme.

- Capillaries lie very close to the **inner capsular membrane** (see above).
- The capillaries have many tiny pores.
- The capsular membrane consists of **podocytes**.
- **Podocytes** help to support the basement membrane of the capillaries.
- It is the basement membrane that is the selective **high-pressure sieve**.
- Only molecules which are small enough can pass through.

Reabsorption

Capillaries from the glomerulus extend to a network across both proximal and distal tubules. The close contact between capillary and tubule is important.

Cross section through proximal tubule

Remember what reabsorption is – the return of substances into the blood which have just left.

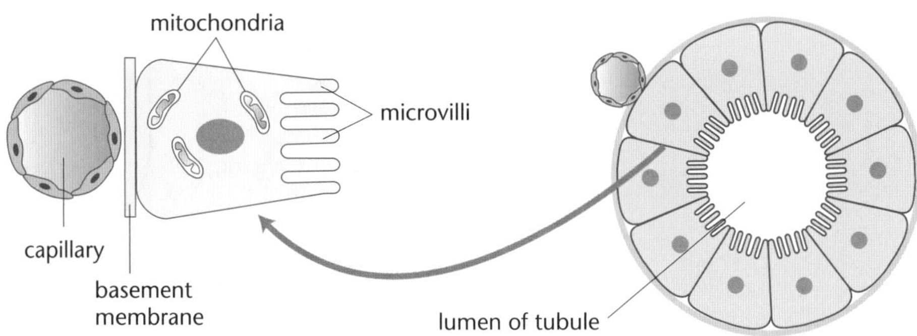

- Substances, such as **glucose**, **urea**, and **water** travel along the tubule.
- Each tubule is one cell thick, consisting of epithelial cells with **microvilli** on the outer membrane.
- Microvilli give a **large surface area** to allow the efficient transport of substances to cross to the capillaries.
- **Carrier proteins** on the microvilli **reabsorb** glucose from the filtrate into the tubules.
- Glucose molecules are then actively transported into the fluids surrounding the capillaries.
- Glucose molecules finally enter the capillaries and so have re-entered the blood.
- By the end of the proximal tubule **all glucose** has been **returned to the blood**.

The distal tubule is also in close contact with the capillary network. Even more reabsorption can take place here.

How do the kidneys conserve water?

Water molecules which pass into the tubule and reach the kidney pelvis continue down a ureter and are lost in urine. Such water loss is carefully controlled; some is always reabsorbed. This control involves both the nervous system, the endocrine system and structures along a nephron. The diagram below outlines the role of the **countercurrent multiplier** in the control of water content in the body.

Countercurrent multiplier

The vasa recta capillaries follow the path of the loop of Henlé to:

(a) supply oxygen to the cells so that active transport of Na^+, and Cl^- can take place efficiently (the process needs energy!)
(b) remove CO_2
(c) reabsorb water.

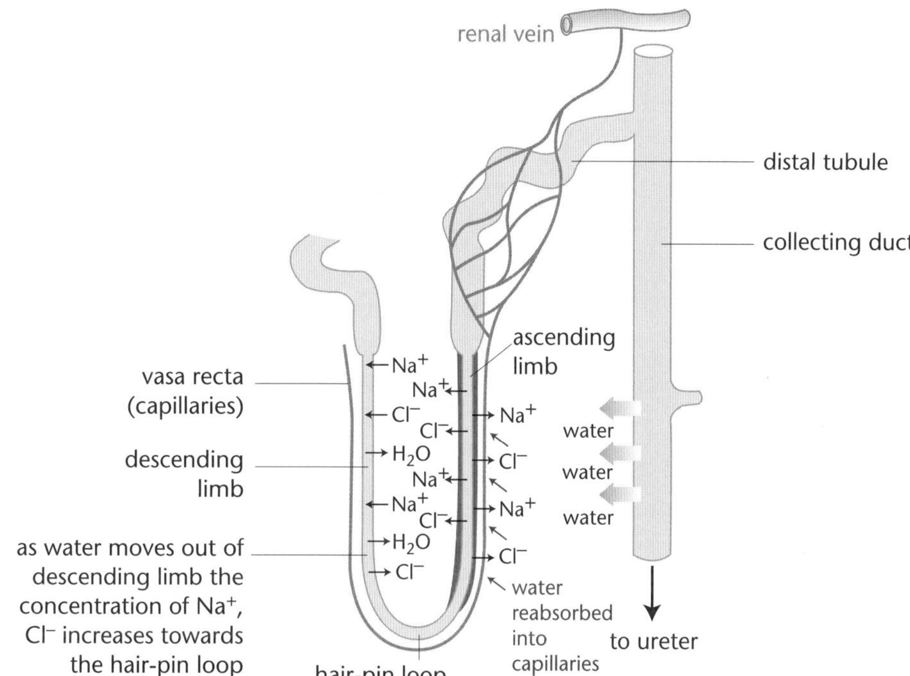

The role of the loop of Henlé

- Na^+ and Cl^- ions are actively transported into the medulla from the ascending limb of the loop of Henlé.
- The ascending limb is thicker than the descending one, and impermeable to the outward movement of water so only the ions leave.
- The Na^+ and Cl^- ions slowly diffuse into the descending limb resulting in their greater concentration towards the base of the loop.
- A high concentration of Na^+ and Cl^- ions in the medulla causes water to leave the collecting duct by osmosis.
- Additionally, water leaves the descending limb by osmosis due to the ions in the medulla.
- Water molecules pass into the capillary network and are successfully reabsorbed.

The role of the distal tubule

The distal tubule is also a site of more reabsorption. Even more substances are returned to the blood here.

The structure of the distal tubule is similar to the proximal tubule, however its specific roles are:

- maintenance of a constant blood plasma pH at around 7.4
- if blood plasma falls **below** a pH of 7.4 then ionic movements take place
  (**H^+** ions) plasma → filtrate
  (**HCO_3^-** ions) filtrate → plasma
- if blood plasma **rises** above a pH of 7.4 then more ion movements take place
  (**OH^-** ions) plasma → filtrate
  (**HCO_3^-** ions) plasma → filtrate

The control of water balance

It is necessary to control the amount of water in the blood. The kidneys can help to achieve this with their ability to intercept water before it can reach the ureters. There are, however, problems to overcome.

Here the consequences of the two extremes of hot and cold are explained. Do remember that there are a range of conditions **between** these extremes. ADH level changes in response to osmoreceptor sensory input to the hypothalamus.

> **KEY POINT**
>
> In hot conditions we lose a lot of water by sweating; too much loss would lead to dehydration problems.
>
> In cold conditions much less water is lost by sweating, giving a potential problem of too much water being retained in the blood.
>
> A balance must be achieved!

Hormonal control of the kidneys – the role of ADH

Control is achieved with the help of antidiuretic hormone (ADH), produced by the hypothalamus and secreted by the posterior lobe of the pituitary gland.

Look out for graphs in questions about kidneys.

- Levels of key substances may be shown.
- If water content down a collecting duct decreases as water content in the medullary region increases:
  - then water molecules are crossing the collecting duct
  - sodium and chloride ions have drawn this water from the collecting duct into the medulla by osmosis.

Scenario 1: warm environmental conditions

- **Osmoreceptors** in the hypothalamus detect an **increase** in the solute concentration of the blood plasma.
- The **hypothalamus** then produces, by neurosecretion, the hormone **ADH**.
- The ADH is secreted into the posterior lobe of the **pituitary gland**.
- From here it passes into the blood and finally reaches the target organs, the **kidneys**.
- Here ADH **increases** permeability of:
  - (i) the collecting ducts
  - (ii) the distal tubules.
- The effect is that more water can be **reabsorbed** back into blood.

The events outlined above give a maximum effect of the countercurrent multiplier. Too much water would be lost by sweating so the water component of the urine must be drastically limited. The resulting urine is therefore low in water content and high in solutes.

Scenario 2: cold environmental conditions

- **Osmoreceptors** in the hypothalamus detect a **decrease** in the solute concentration of the blood plasma.
- The hypothalamus then produces **less ADH**.
- Less ADH leaves the posterior lobe of the pituitary gland.
- Less ADH reaches the target organs, the kidneys.
- The collecting ducts and the distal tubules are **not so permeable**.
- **Less water** can be **reabsorbed** back.

The urine is of greater volume due to greater water content. No wonder we urinate more in cold weather!

Renal dialysis

Anticoagulant is added to stop the blood clotting whilst it is in the machine.

People may have kidney failure for a number of reasons. This may require them to have renal dialysis. This involves linking the person up to a dialysis machine at regular intervals. Blood is removed from a vein in the arm. It then passes through a long coiled tube made of partially permeable cellophane. The fluid surrounding the tube contains water, salts, glucose and amino acids but no waste products such as urea. These waste materials therefore diffuse out of the blood into the fluid.

Sample question and model answer

Note the close proximity of cell A to the capillary. This gives a clue as to its function.

The diagram shows a cell of the inner wall of a renal (Bowman's) capsule. The two structures shown in the diagram are very important in the passage of substances out of the blood into the proximal tubule.

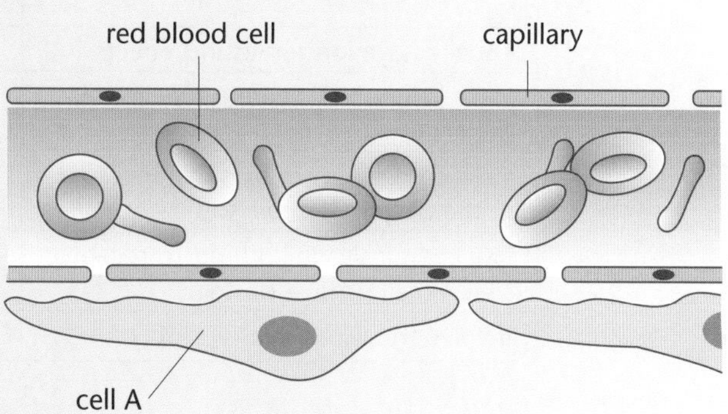

(a) Name cell A. [1]

podocyte

(b) Explain how the cells of the inner capsule wall and the capillaries of the glomerulus help in the process of ultrafiltration. [5]

The question shows that five marks are available. Make sure that you give at least five points to gain your marks. Superficial answers fall short of the total.

Capillaries lie very close to the inner capsular wall; the capillaries have pores; the podocytes are shaped so that many gaps exist between the capsular wall and capillaries; the podocytes help to support the capillary basement membrane which is under pressure; only molecules which are small enough are forced through the basement membrane so the process is selective.

(c) As glomerular filtrate leaves the renal (Bowman's) capsule it enters the proximal convoluted tubule. The graph below shows the ratio of glucose and urea in the blood plasma and the filtrate through the proximal tubule. A ratio of 1.0 means that the concentration in both plasma and filtrate are the same.

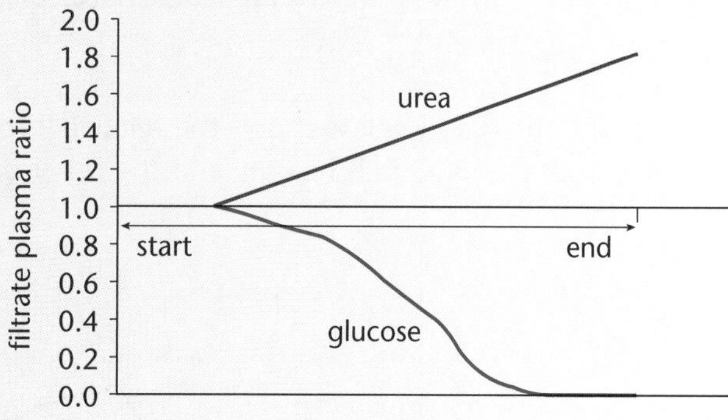

You may find this difficult. However, you can link the fact that the kidney nephron does reabsorb useful substances but not the waste, urea. Relate this to the graph then your task is possible!

Explain the changes in plasma–filtrate from the beginning to the end of the proximal tubule for:

(i) glucose [3]

As the fluid moves along the tubule there is increasingly more glucose in the plasma than the filtrate; this is because glucose is reabsorbed into the blood; all glucose is returned to the blood before end of proximal tubule.

(ii) urea. [3]

As the fluid moves along the tubule there is increasingly more urea in the filtrate than in the blood plasma; no urea is reabsorbed so it remains in the tubule; water is reabsorbed which has the effect of increasing the relative concentration of urea.

Practice examination questions

1 (a) Complete the table below to compare the nervous and endocrine systems. Put a tick in each correct box for the features shown.

	Nervous system	*Endocrine system*
Usually have longer lasting effects		
Have cells which secrete transmitter molecules		
Cells communicate by substances in the blood plasma		
Use chemicals which bind to receptor sites in cell surface proteins		
Involve the use of Na^+ and K^+ pumps		

[2]

(b) Name the process which keeps the human body temperature and water content of blood regulated. [1]

[Total: 3]

2 A mammal is in hot environmental conditions. Explain the effect of a high quantity of ADH entering the blood from the pituitary gland. [6]

3 (a) The products of transamination are represented below. Complete the equation.

→ C_2H_5 (amino acid) + CH_3 (keto acid) [2]

(b) (i) Where in the human body does transamination take place? [1]

(ii) Why is transamination necessary in the human body? [2]

[Total: 5]

4 The graph below shows the relative levels of glucose in the blood of two people: A and B. One is healthy and the other one is diabetic.

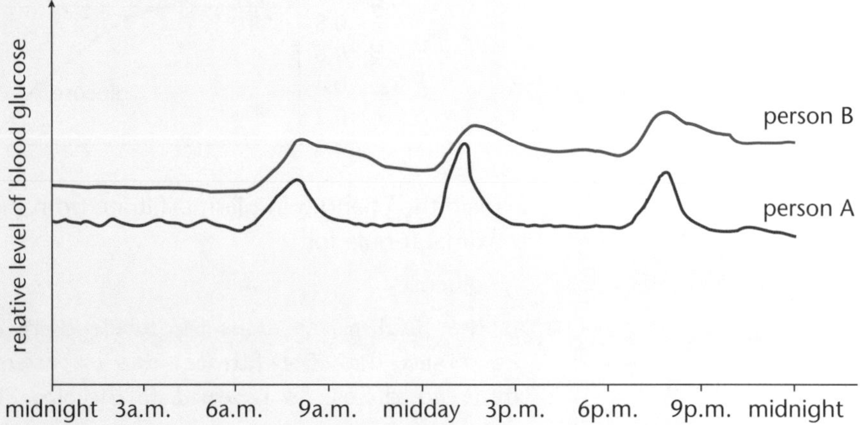

(a) Which person is diabetic? Give evidence from the graph for your answer. [1]

(b) What is the evidence that both people produce insulin? [1]

(c) Where in the body is insulin produced? [2]

[Total: 4]

Chapter 4

Further genetics

The following topics are covered in this chapter:

- *Genes, alleles and protein synthesis*
- *Inheritance*
- *Cell division*

4.1 Genes, alleles and protein synthesis

After studying this section you should be able to:

- *explain how proteins are produced in cells*
- *describe various methods by which gene expression is controlled*
- *define a range of important genetic terms*

LEARNING SUMMARY

Genes, alleles and protein synthesis

OCR 5.1.1

A gene is a section of DNA which controls the production of a polypeptide in an organism. The total effects of all of the genes of an organism are responsible for the characteristics of that organism. Each protein contributes to these characteristics whatever its role, e.g. structural, enzymic or hormonal.

The order of bases in the gene is called the genetic code and will code for the order of amino acids in the polypeptide. The order of amino acids is called the primary structure of the protein and will determine how the protein folds up to form the secondary and tertiary structures. The formation of a protein molecule is called protein synthesis.

Protein synthesis

The process of protein synthesis involves the DNA and several other molecules.

- **Messenger RNA:** This is a single-stranded nucleotide chain that is made in the nucleus. It carries the complementary DNA code out of the nucleus to the ribosomes in the cytoplasm.
- **mRNA polymerase:** This is the enzyme that joins the mRNA nucleotides together to form a chain.
- **ATP:** This is needed to provide the energy to make the mRNA molecule and to join the amino acids together.
- **tRNA:** This is a short length of RNA that is shaped rather like a clover leaf. There is one type of tRNA molecule for every different amino acid. The tRNA molecule has three unpaired bases that can bind with mRNA on one end and a binding site for a specific amino acid on the other end.

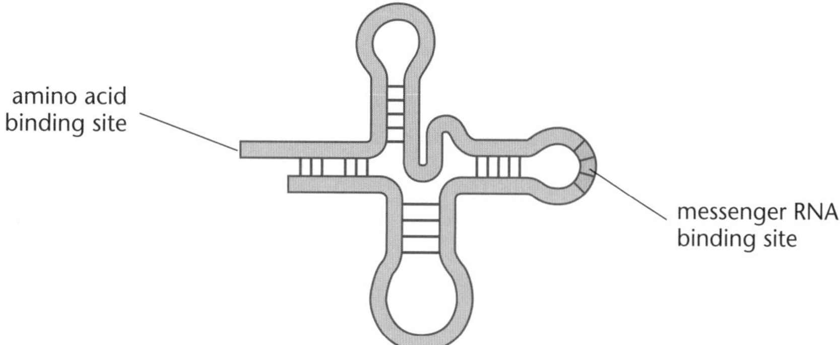

Key points from AS

- **The genetic code**
  Revise AS pages 72–74

The following diagrams show protein synthesis.

1 In the nucleus **RNA polymerase** links to a start code along a DNA strand.

2 RNA polymerase moves along the DNA. For every organic base it meets along the DNA a complementary base is linked to form mRNA (**messenger RNA**).

There is no thymine in mRNA. Instead there is another base, uracil.

Pairing of organic bases				
DNA	G	C	T	A
mRNA	C	G	A	U

3 RNA polymerase links to a stop code along the DNA and finally the mRNA **moves** to a **ribosome**. The DNA stays in the nucleus for the next time it is needed.

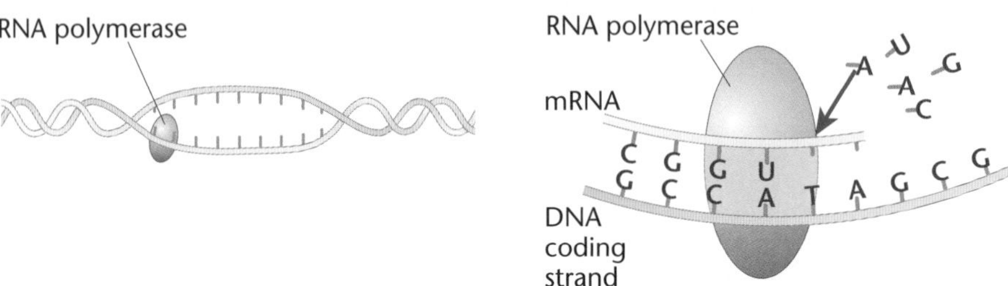

The transfer of the code from DNA to mRNA is called transcription.

4 Every three bases along the mRNA make up one **codon** which codes for a specific amino acid. Three complementary bases form an **anticodon** attached to one end of tRNA (**transfer RNA**). At the other end of the RNA is a specific amino acid.

5 All along the mRNA the tRNA 'partner' molecules enable each amino acid to bond to the next. A chain of amino acids (**polypeptide**) is made, ready for release into the cell.

Note the link between each pair of amino acids along a polypeptide – the peptide link.

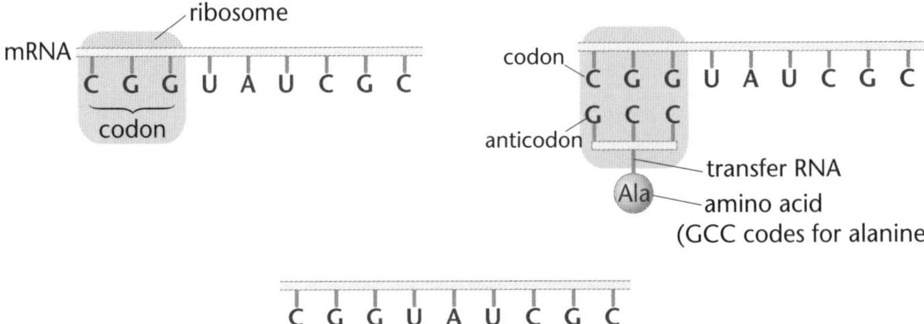

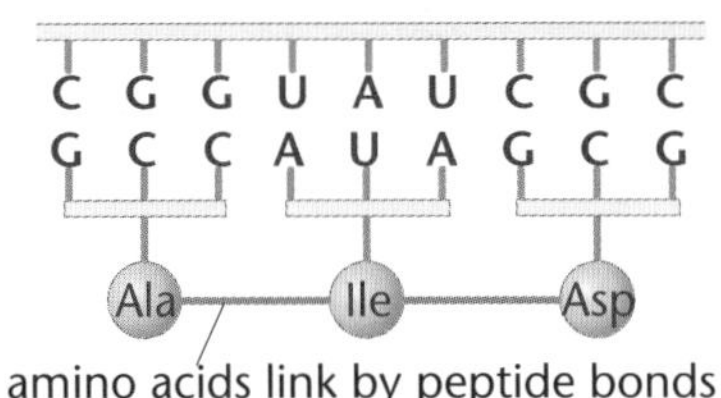

The conversion of the mRNA code to a sequence of amino acids is called translation.

Control of protein synthesis

OCR 5.1.1

In a multicellular organism, every cell contains all the genetic material needed to make every protein that the organism requires. However, as they develop, cells become specialised. This means that they do not need to use every protein and so it would be a waste to make every protein all the time. Genes must be switched on and switched off.

Most of the early work on gene regulation was carried out on bacteria which are prokaryotic.

Jacob and Monod's theory

During the 1950s, Jacob and Monod found that the bacterium *E.coli* would only produce the enzyme lactase if lactose was present in the growth medium. The production of lactose was controlled by three different genes:

- a structural gene codes for the enzyme
- an operator gene which turns the structural gene on
- a regulator gene that produces a chemical that usually stops the action of the operator gene.

If lactose is present, the action of the chemical inhibitor is blocked and lactase is made.

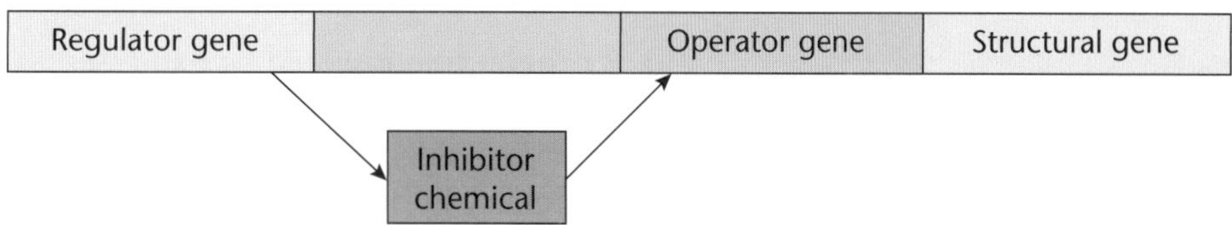

KEY POINT

The combination of the three genes controlling lactase production is called the ***lac* operon.**

Gene control in eukaryotes

In mature plants, many cells remain totipotent but in mature animals these totipotent or **stem cells** are harder to find.

In eukaryotic cells, gene regulation seems to be much more complicated. Cells in the early embryo are called **totipotent**. This means that they can develop into any type of cell. These cells produce all the cells of a multicellular organism and the specialised cells have to be produced in the correct place.

Scientists are trying to work out how this is done and have found genes called **homeobox genes**. These genes seem to produce proteins that act as transcription factors turning on other genes. Similar homeobox genes have been found in animals, plants and fungi.

There is much interest at present in the possible use of siRNA to treat various genetic conditions.

Other factors from the cytoplasm can also effect transcription. **Steroid hormones** such as oestrogen can bind with **receptors** in the cytoplasm and then move into the nucleus causing genes to be transcribed.

Scientists have recently found a different type of RNA. This is a small double-stranded molecule called siRNA (small interfering RNA). This seems to silence the action of certain genes.

Essential genetic terms

OCR 5.1.2

It is necessary to understand the following range of specialist terms used in genetics.

Allele – an alternative form of a gene, always located on the same position along a chromosome. This position is called a locus.

E.g. an allele coding for the white colour of petals

Dominant allele – if an organism has two different alleles then this is the one which is expressed, often represented by a capital letter.

E.g. an allele coding for the red colour pigment of petals, **R**

Check out all of these genetic terms.

- Look carefully at the technique of giving an example with each definition. Often examples help to clarify your answer and are usually accepted by the examiners.
- In examination papers you will need to apply your understanding to **new** situations.
- Genetics has a specialist language which you will need to use.

Recessive allele – if an organism has two different alleles then this is the one which is **not** expressed, often represented by a lower case letter. Recessive alleles are only expressed when they are not masked by the presence of a dominant allele.

E.g. an allele coding for the white colour pigment of petals, **r**.

Homozygous – refers to the fact that in a diploid organism both alleles for a particular gene are the same.

E.g. **R R** or **r r**.

Heterozygous – refers to the fact that in a diploid organism both alleles for a particular gene are different.

E.g. **R r** (petal colour would be expressed as red).

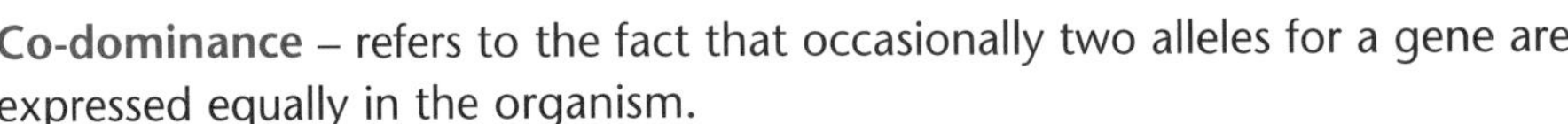

Co-dominance – refers to the fact that occasionally two alleles for a gene are expressed equally in the organism.

E.g. **A**, **B** alleles = **AB** (blood group with antigens A and B).

A number of inherited alleles of a range of genes often exhibit continuous variation, e.g. height. Each allele contributes small incremental differences. That is why there are smooth changes in height across a population.

Polygenic inheritance – where an inherited feature is controlled by two or more genes, along different loci along a chromosome. This results in continuous variation.

E.g. the height of a person is controlled by a number of different genes.

Remember that both sperms and ova are haploid.

Haploid – refers to a cell which has a single set of chromosomes.
E.g. a nucleus in a human sperm has 23 single chromosomes.

Diploid – refers to a cell which has two sets of chromosomes.
E.g. a nucleus in a human liver cell has 23 pairs of chromosomes.

In diploid cells one set of chromosomes is from the male parent and one from the female.

Homologous chromosomes – refers to the pairs of chromosomes seen during cell division. These chromosomes lie side by side, each gene at each locus being the same.

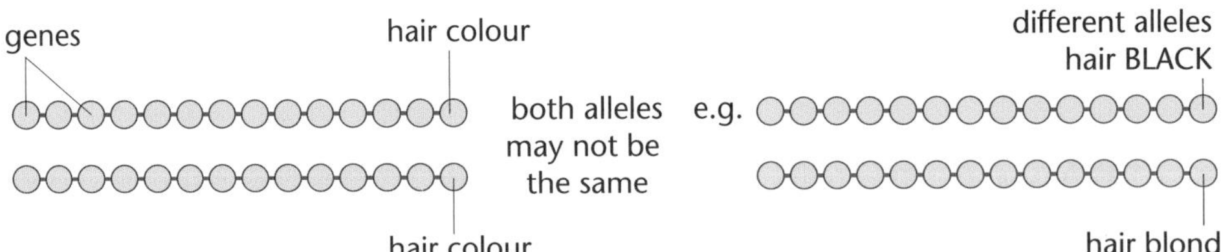

Often polyploid organisms cannot reproduce sexually but asexually they are successful.

Polyploid – refers to the fact that a cell has three or more sets of chromosomes. This can increase yield.

E.g. cultivated potato plants are **tetraploid**, that is four sets of chromosomes in a cell. (Tetraploidy is a form of polyploidy.)

Genotype – refers to all of the genes found in the nuclei of an organism, including both dominant and recessive alleles.

dominant
E.g. A B c d E F g H i (all alleles are included in a genotype)
a B C D e f g h I
recessive

phenotype = genotype + environment

Phenotype includes all alleles which are expressed in an organism. The environment supplies resources and conditions for development. Varying conditions result in an organism developing differently. Identical twins fed different diets will show some differences, e.g. weight.

Environment has considerable effect.

Linkage – refers to two or more genes which are located on the same chromosome.

E.g. linked X------------------Y-- not linked X-------------------- ← different
same chromosomes ↗ ------------------Y-- ← chromosomes

Somatic cell – refers to any body cell which is not involved in reproduction.
E.g. liver cell

Autosome – refers to every chromosome apart from the sex chromosomes, X and Y.

4.2 Cell division

After studying this section you should be able to:

- *compare the main features of mitosis and meiosis*
- *describe and explain the process of meiosis*
- *understand the consequences of chiasmata (crossing over)*

LEARNING SUMMARY

Why are there two types of cell division?

OCR 5.1.2

Each type of cell division has a different purpose.

Mitosis

There are occasions when it is necessary to **replicate** cells, e.g. for growth and repair. This is the role of **mitosis**. It produces a clonal line of cells. Each cell divides to form **two diploid** cells, identical in every way.

At AS Level you learned the names of the stages in sequence. The stages of meiosis use the same names, in the same order but there are two nuclear divisions this time!

Meiosis

This is needed in gamete formation. In human cells a body (somatic) cell has 46 chromosomes. If each gamete contained 46 chromosomes then the zygote produced at fertilisation would have 92 chromosomes. This would be lethal! Meiosis is also called **reduction division** because the gametes produced are haploid. In human gametes the haploid chromosome number is 23. Each cell divides to form **4 haploid** cells. Every cell is different to the parent cell and each other.

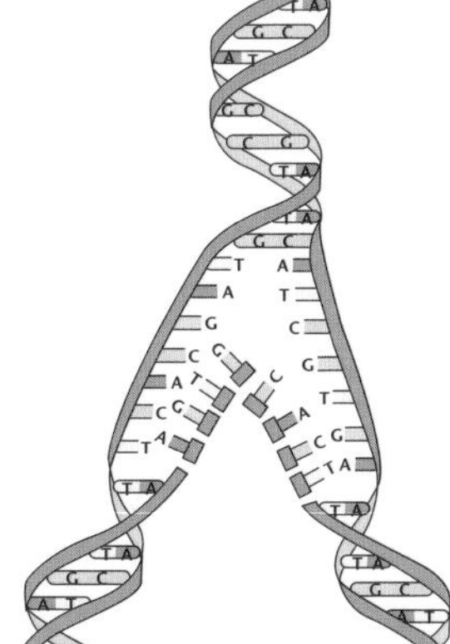

Meiosis: the process explained

The preparation of a cell prior to meiotic division is during **interphase**. During this pre-stage each double strand of DNA replicates to produce **two** exact copies of itself. This also takes place in exactly the same way before mitosis takes place. After interphase, when the cell division commences, major differences occur.

In meiosis during the first stage, **prophase 1**, a fundamentally important event takes place, where chromatids **cross over**. Each crossover is termed a **chiasma**.

The mechanism of crossovers (chiasmata)

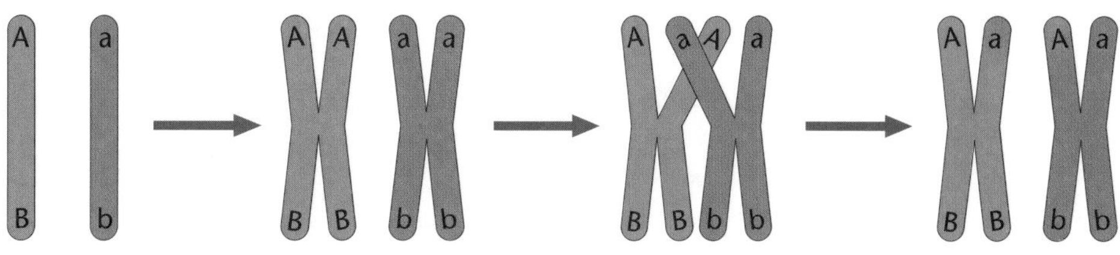

Key points from AS

- **Cell division**
  Revise AS pages 75–76

A represents an allele dominant to **a**, a recessive allele.

B represents an allele dominant to **b**, a recessive allele.

In humans, with **many chiasmata** taking place along **all 23 pairs of chromosomes**, every cell at the completion of meiosis is genetically different.

> Chiasmata result in **different allele combinations**! — KEY POINT

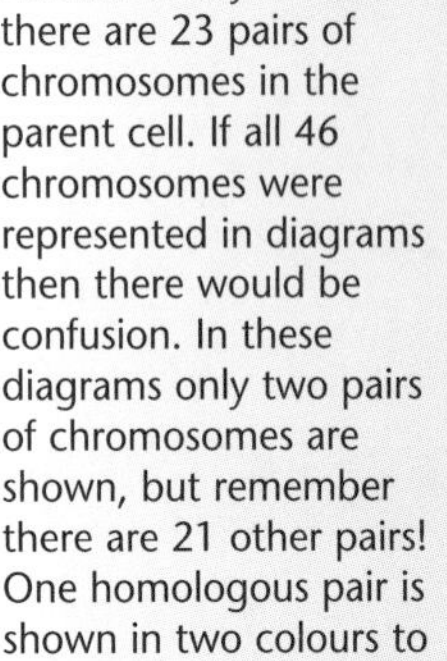
In the division of a human cell by meiosis there are 23 pairs of chromosomes in the parent cell. If all 46 chromosomes were represented in diagrams then there would be confusion. In these diagrams only two pairs of chromosomes are shown, but remember there are 21 other pairs! One homologous pair is shown in two colours to show the consequence of crossovers.

The process of meiosis

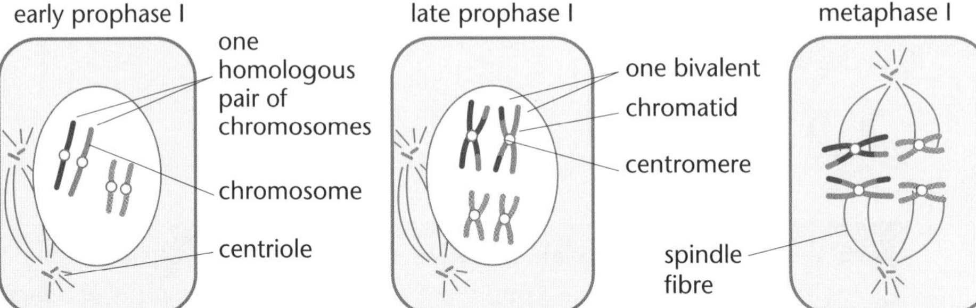

each chromosome forms two chromatids; two centrioles begin to move forming a spindle

the chromatids have crossed over and exchanged DNA at two positions

bivalents from homologous chromosomes lie parallel to each other along the equator

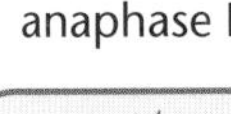

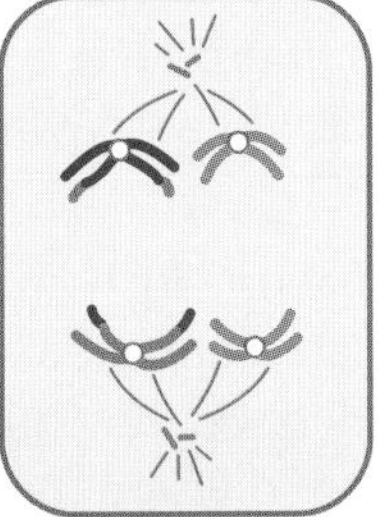

corresponding bivalents are pulled by spindle fibres towards poles

telophase I

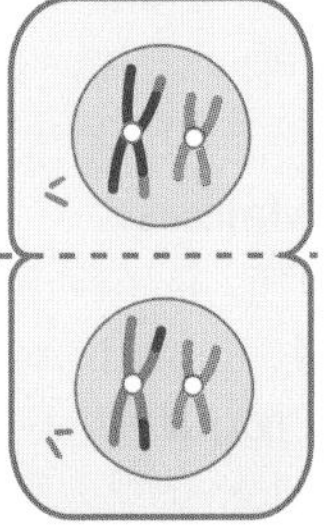

the cell constricts at the equator to form two cells

prophase II

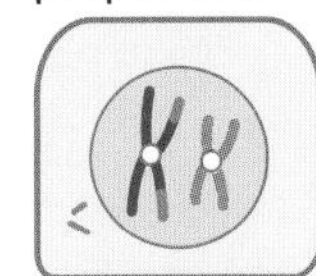

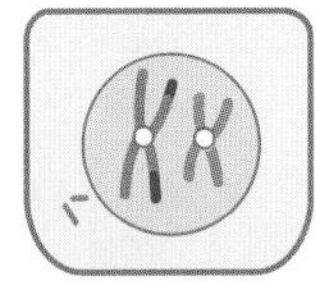

EACH of these cells will divide to form two cells

metaphase II

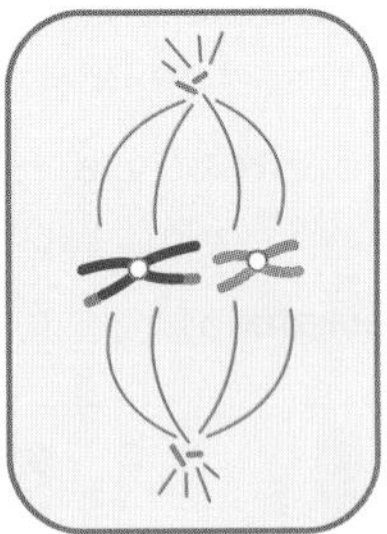

single bivalents lie across the equator

anaphase II

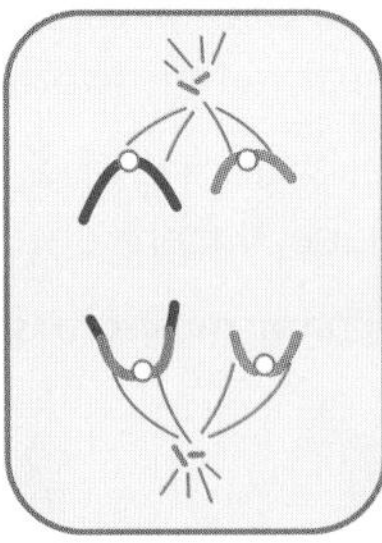

spindle fibres contract to pull the centromere apart; single chromosomes are dragged to the poles

early telophase II

cell constricts at equator

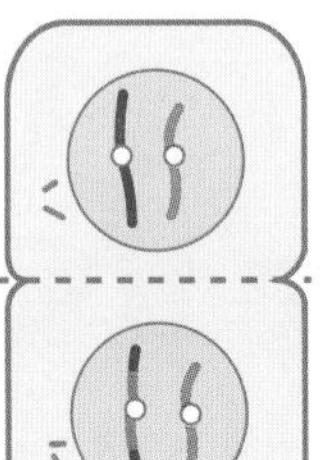

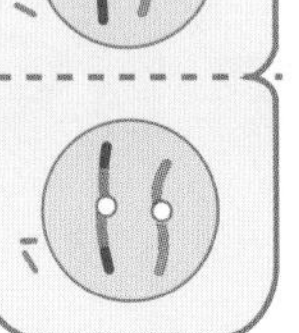

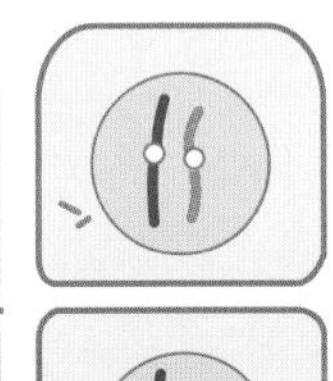

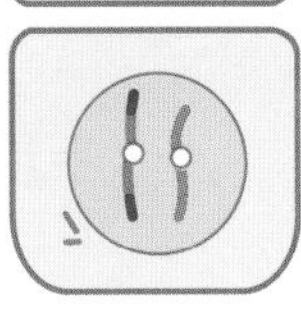

two cells are produced for each cell from first division = four daughter cells

The significance of meiosis

This diagram shows the single chromosomes produced as a result of two crossovers.

In an examination you will need to understand the consequence of many crossovers. Crossovers are a source of genetic variation.

end of first meiotic division

'sister' chromatids still attached

A a / B B   A a / b b

end of second meiotic division

A B   a B   A b   a b

'sister' chromatids have parted from the centromere

The combination of each of these chromosomes with others results in further genetic variation.

A represents an allele dominant to **a**, a recessive allele.

B represents an allele dominant to **b**, a recessive allele.

- Many more than two crossovers can take place between each homologous pair. The presence of 23 homologous pairs of chromosomes in a diploid human cell results in a lot of crossovers.
- Once the chromatids finally separate in **anaphase II**, each moves with 22 others to a pole to produce a daughter cell.
- After division, four different chromatids are produced from each homologous pair (see above).

1 from 4 chromatids combine with 1 from another 4 chromatids. These combinations give **16 possibilities.** Add the combination of another 1 from 4 chromatids and there are **64 possibilities**. Another 1 from 4 is added ... and another ... to include all 23 pairs. This gives millions of combinations. No wonder we all look different!

KEY POINT

What determines which of the four chromatids of one homologous pair is grouped with chromatids from the other homologous pairs?

The answer is '**chance**' and the combination of the 23 single chromosomes dragged through the cytoplasm by the spindle fibres, is known as **independent assortment**.

Every gamete produced by meiosis is genetically different. However, there are two sexes. This means that in sexual reproduction, the fact that there are two different gametes which combine their alleles in the zygote, gives another major source of variation.

Progress check

The diagram below shows a stage in cell division.

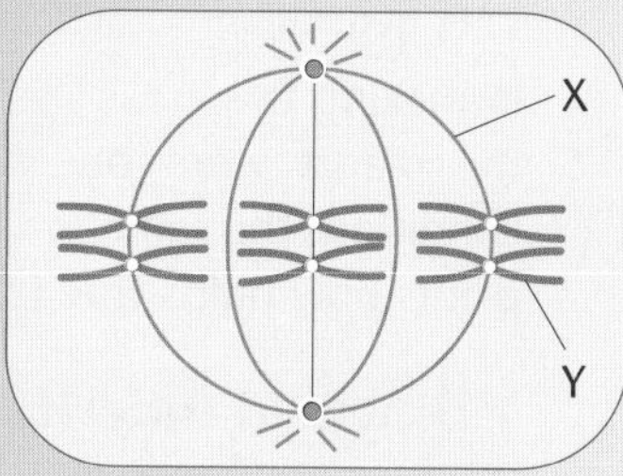

(a) Name parts X and Y.

(b) Which type of cell division is shown? Give a reason for your answer.

(a) X = spindle fibre Y = chromatid or bivalent

(b) Metaphase I of meiosis. Bivalents line up in twos along the equator whereas in mitosis they lie singly.

4.3 Inheritance

After studying this section you should be able to:

LEARNING SUMMARY

- *understand Mendel's laws of inheritance*
- *understand the principles of monohybrid inheritance and dihybrid inheritance*
- *understand exceptions to Mendel's laws such as linkage, sex linkage, co-dominance and epistasis*
- *describe the principle of sex determination*
- *use the Hardy–Weinberg principle to predict the numbers of future genotypes*
- *use chi-squared to test actual genetic data against a predicted ratio*

Mendel and the laws of inheritance

OCR 5.1.2

Gregor Mendel was the monk who gave us our understanding of genetics. He worked with pea plants to work out genetic relationships.

> KEY POINT
>
> Mendel's first law indicates that:
> - each character of a diploid organism is controlled by a pair of alleles
> - from this pair of alleles only one can be represented in a gamete.

Monohybrid inheritance

Always show your working out of a genetical relationship in a logical way, just like solving a mathematics problem.

Mendel found that when homozygous pea plants were crossed, a predictable ratio resulted. The cross below shows Mendel's principle.

pea plants	pea plants
T = TALL (dominant)	t = dwarf (recessive)

A homozygous TALL plant was crossed with a homozygous recessive plant

If you have to choose the symbols to explain genetics, then use something like **N** and **n**. Here the upper and lower cases are very different. **S** and **s** are corrupted as you write quickly and may be confused by the examiner awarding your marks.

TT × tt

gametes (T) (T) (t) (t)

F_1 generation Tt

All offspring 100% TALL, and heterozygous.

Heterozygous plants were crossed

Tt × **Tt**

gametes (T) (t) (T) (t)

	(T)	(t)
(T)	TT	Tt
(t)	Tt	tt

2 × 2 Punnet square to work out different genotypes

F_2 generation 3 TALL : 1 DWARF

In examinations you may have to work out a probability. 3:1 is the same as a 1 in 4 chance. Remember only large numbers would confirm the ratio.

In making this cross Mendel investigated **one** gene only. The height differences of the plants were due to the different alleles. Mendel kept all environmental conditions the same for all seedlings as they developed. The 3:1 ratio of tall to short plants only holds true for large numbers of offspring.

Dihybrid inheritance

Mendel used pea plants to again work out the genetic relationship between plants for genes at different loci (positions) on chromosomes.

> KEY POINT
>
> Mendel's second law of independent assortment indicates that:
>
> either of a pair of alleles, say **A** and **a**, can combine with either of another pair, say **B** and **b**.

You probably will not be given the example of pea plants in your examinations. Coloured petals or fruit fly features are often given. Make sure that you clearly show the symbols and apply the principles learned from this example. The key part in a genetics question is that you give correct gametes.

For a heterozygous genotype of AaBb the gametes are

AB Ab aB ab

NOT

~~AA~~ ~~aa~~ ~~BB~~ ~~bb~~

The cross below shows Mendel's dihybrid principle.

pea plants	*pea plants*
R = round seeds (dominant)	r = wrinkled seeds (recessive)
Y = yellow seeds (dominant)	y = green seeds (recessive)

A homozygous dominant plant with yellow, round seeds was crossed with a homozygous recessive plant with green, wrinkled seeds.

RRYY × rryy

gametes (RY) (ry)

F_1 generation RrYy

seeds 100% round, yellow, and heterozygous

Below, heterozygous plants from F_1 generation were crossed.

Be ready to cross two organisms from a punnet square, e.g.

RrYY × rrYy

Do not expect a 9:3:3:1 ratio!

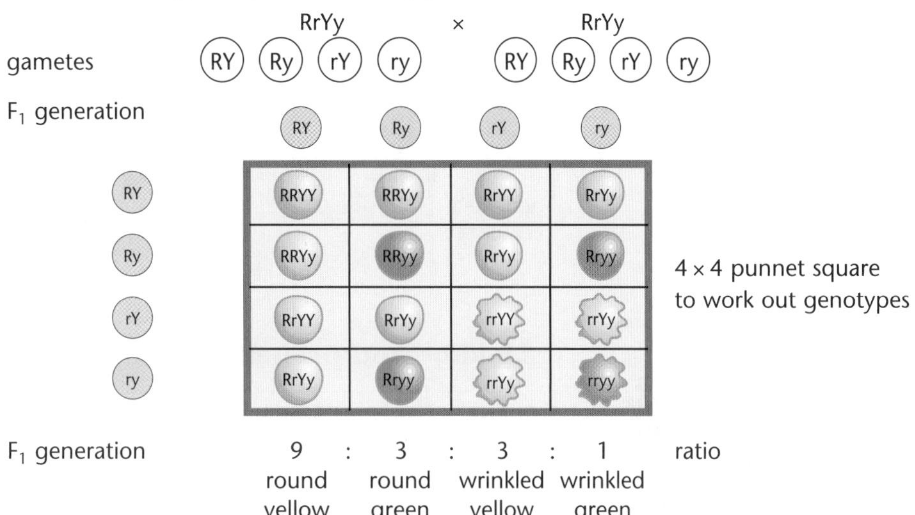

The 9 : 3 : 3 : 1 ratio only holds true for large numbers of offspring.

The principles above can be applied to any dihybrid example. The F_1 generation is so predictable that many varieties of commercial crop are grown from F_1 generation seeds, known as F_1 hybrids.

Linkage

OCR 6.1.2

Mendel was rather lucky in the characteristics that he chose. If he had chosen other characteristics there are a number of complications that may have prevented him drawing the correct conclusions. One of these complications is linkage.

Each chromosome consists of a sequence of genes. All genes along a chromosome are **linked** because they are part of the same chromosome. Most chromosomes have between 500 and 1000 genes in a linear sequence. These genes are linked.

Linkage can occur on autosomes and sex chromosomes.

What is the significance of linkage?

We are able to make predictions about the proportion of future offspring when we know the genotype of parents, like the 9:3:3:1 ratio for dihybrid inheritance. This is only true if the pair of contrasting genes are **on different chromosomes.** Consider these two alternatives:

A dominant, **a** recessive; **B** dominant, **b** recessive

loci (positions) of genes

A a B b

AaBb × **AaBb**

Not linked This cross would produce 9:3:3:1 proportion in offspring

A a B b

AaBb × **AaBb**

Linked This cross would be unlikely to produce a 9:3:3:1 proportion in offspring

When two genes, e.g. A a and B b are on **different chromosomes** then their inheritance together is **not affected by crossovers**. Either of one pair **can** be inherited with either of the other pair. The relationship changes when the genes are along the same chromosome. Crossovers are affected! Alleles along the same gene locus can be swapped from one chromatid to another.

Consider these alternatives for linked genes, where a homozygous dominant genotype is crossed with a homozygous recessive.

The more crossovers there are, the greater the chance that the four different gene combinations will be produced in each parental genotype. They could produce a 9:3:3:1 proportion in cross 1. However, if the genes are closer along the chromosome then the proportions deviate significantly from this pattern. Genes **adjacent** to each other tend to be **inherited together**, because the chance of them being parted is very low.

Crossover 1 (AABB × aabb)

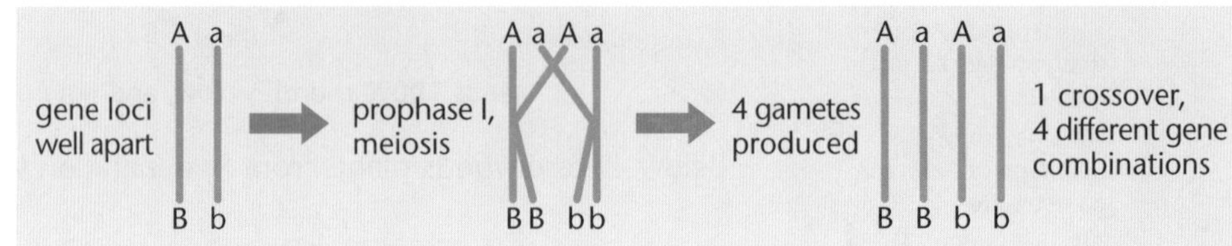

Crossover 2 (AABB × aabb)

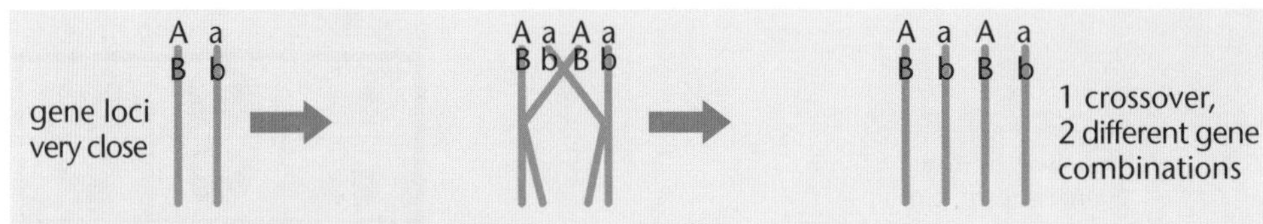

Linkage and probability

Consider these crosses

Cross one

R = red petals (dominant)   r = blue (recessive)
L = long stems (dominant)   l = short stems (recessive)
homozygous red petals long stemmed   homozygous blue petals short stemmed

RRLL × rrll

The young organisms are often termed progeny or offspring.

gametes: R L, R L

F_1 generation   RrLl   100% heterozygous red petals, long stemmed

Cross two

Heterozygous red petals long stemmed   Heterozygous red petals long stemmed

RrLl × RrLl

gametes

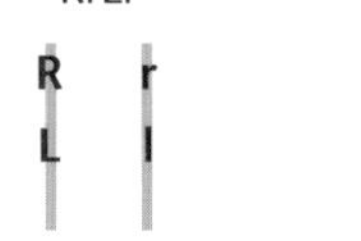

R L, r l

F_2 generation

	(RL)	(rl)
(RL)	RRLL red long	RrLl red long
(rl)	RrLl red long	rrll blue short

probability is: red petals long stems 3:1 blue petals short stems

actual numbers 610   202

(That is almost the one in four chance.)

Examiner's tip

Do not fall into the 'linkage' trap. Here the 9:3:3:1 ratio is replaced by a 3:1 ratio due to genes being adjacent on the chromosome. The net result is that **RL** and **rl** tend to be inherited together as a package. Do not dive into questions like this. Keep alert!

This example shows the consequence of **very close linkage**. In this genuine example the genes were so close that the RL and rl combinations were never parted by crossovers. No Rl or rL allele combinations were evident. So the classic RrLl × RrLl ratio of 9:3:3:1 was not possible. Instead a 3:1 ratio was produced. This is **not** monohybrid inheritance.

Progress check

(a) List the gametes for the following dihybrid cross.
(The genes are not linked.)
Ddee × DDEe

(b) Show the genotypes of the progeny.

(a) De de DE De
(b) DDEe, DdEe, DDee, Ddee

Sex determination and sex linkage

OCR 5.1.2

The genetic information for gender is carried on specific chromosomes. In humans there are 22 pairs of autosomes plus the special sex determining pair, either **XY (male)** or **XX (female)**. In some organisms such as birds this is reversed.

Some genes for sex determination are on autosomes but are activated by genes on the sex chromosomes.

Sperm can carry an X or Y chromosome, whereas an egg carries only an X chromosome.

The genetic cross shown should not give you any problems at A2 Level. However, look out for the combination of another factor which will increase difficulty.

Remember that X and Y are chomosomes and not genes.

genotype: XX (female) × XY (male)

gametes: (X) (X) (X) (Y)

offspring:

	(X)	(X)
(X)	XX	XX
(Y)	XY	XY

probability: 1 male : 1 female

This shows how 50:50 males to females are produced.

Sex linkage

Look more closely at the structure of the X and Y chromosomes.

Why do they not look like X and Y? Only when the cells are dividing, do they take the XY shape, after chromatid formation.

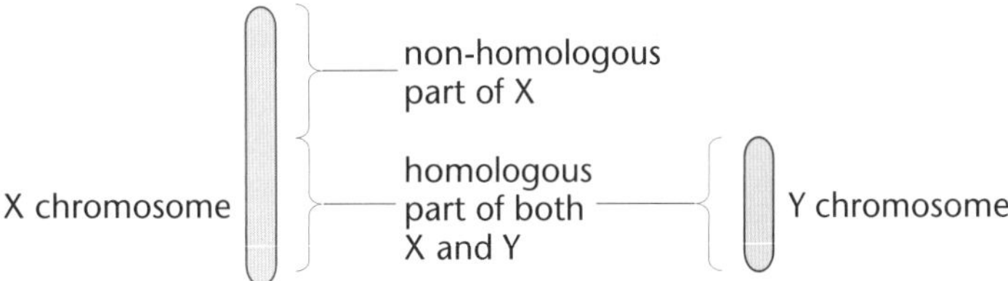

Homologous part of the sex chromosomes

- Has the same genes in both sexes.
- Each gene can be represented by the same or different alleles at each locus.

Non-homologous part of the sex chromosomes

- This means that the X chromosome has genes in this area, whereas the Y chromosome, being shorter, has no corresponding genes.
- Genes in this area of the X chromosome are always expressed, because there is no potential of a dominant allele to mask them.

- There are some notable genes found on the non-homologous part, e.g. haemophilia trait, and colour blindness trait.

The sex chromosomes, X and Y, carry genes other than those involved in sex determination. Examples of such genes are:

- a gene which controls blood clotting, i.e. is responsible for the production of factor VIII vital in the clotting process.
- a gene which controls the ability to detect red and green colours

The loci of both genes are on the non-homologous part of chromosome X.

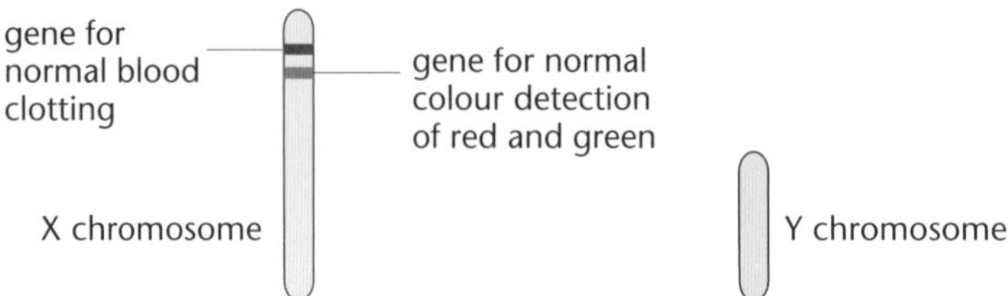

What is the effect of sex linked genes?

The fact that these genes are linked to the X chromosome has no significant effect when the **genes perform their functions correctly**. There are consequences, however, if the genes fail. This can be illustrated by a consideration of **red–green colour blindness**. When a gene is carried on a sex chromosome, the usual way to show this is by X^R.

R = normal colour vision (dominant)   r = red–green colour blindness (recessive)

You can see the four possible genotypes. A female needs two r alleles, a male just needs one.

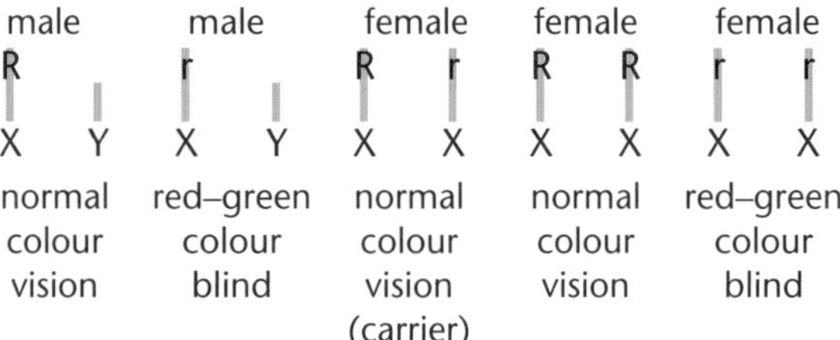

The genetic diagram shows that a female needs two recessive alleles (one from each parent) to be colour blind. A male has only one gene at this locus, so one recessive allele is enough to give colour blindness. The colour blindness gene is rare, so the chances of being a colour blind female are very low.

Consider these crosses.

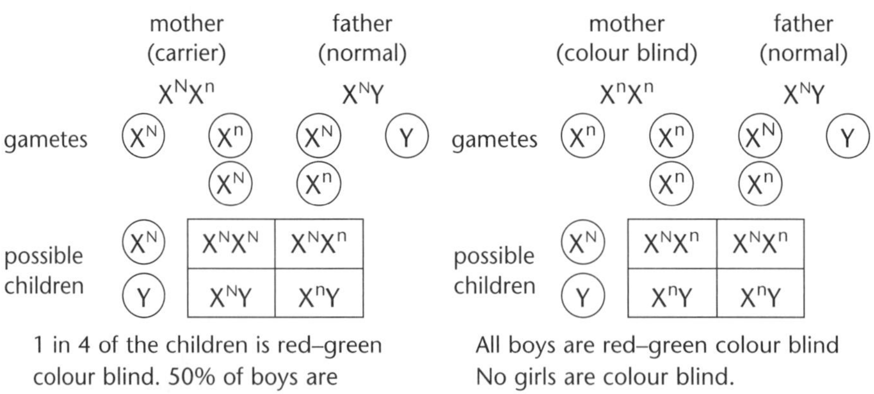

1 in 4 of the children is red–green colour blind. 50% of boys are colour blind, but no girls.

All boys are red–green colour blind
No girls are colour blind.

Progress check

What is the probability of a colour blind male and carrier female producing:

(a) a boy with normal colour vision

(b) a colour blind girl?

Show your working.

mother (carrier) X^NX^n father (colour blind) X^nY

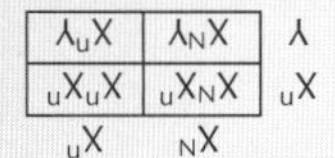

(a) 1 in 4 (b) 1 in 4

Co-dominance

OCR 5.1.2

This term is given when each of two *different* alleles of a gene are expressed in the phenotype of an organism. In humans there are two co-dominant alleles. These alleles produce the antigens in blood which are responsible for our blood groups.

Remember that for co-dominance there is no dominance. Both alleles are equally expressed.

Consider these crosses

Blood group	*Genotypes*
A	I^AI^A, I^AI^O
B	I^BI^B, I^BI^O
AB	I^AI^B
O	I^OI^O

The allele for production of:

A antigen in blood = I^A
B antigen in blood = I^B
No antigen in blood = I^O

Look out for more examples of co-dominance in examination questions, e.g. in shorthorn cattle, R = red and W = white. Where they occur in the phenotype together they produce a dappled intermediary colour known as roan.

Mother × Father
$I^B I^O$ × $I^A I^O$

gametes I^B I^O I^A I^O

children

	I^B	I^O
I^A	I^AI^B	I^AI^O
I^O	I^BI^O	I^OI^O

All blood groups produced by this cross.

Mother × Father
$I^A I^B$ × $I^A I^O$

gametes I^A I^B I^A I^O

children

	I^A	I^B
I^A	I^AI^A	I^AI^B
I^O	I^AI^O	I^BI^O

There must be a I^O from both parents to produce an O blood group.

In this instance, there are two co-dominant alleles, I^A and I^B. When inherited together they are both expressed in the phenotype. Group O blood does not have any antigen.

Epistasis

OCR 5.1.2

This involves two different genes which affect each other. A form of epistasis can be explained by referring to the sweet pea plant. *Lathyrus odoratus* is a white flowered sweet pea. When crossed, two white parent plants can produce white and purple flowers. This can be explained as follows:

Dominant epistasis also exists. In this instance a dominant allele can **inhibit** another, e.g. in the land snail: **A** a dominant allele inhibits **B, b** alleles responsible for banding on the shell.

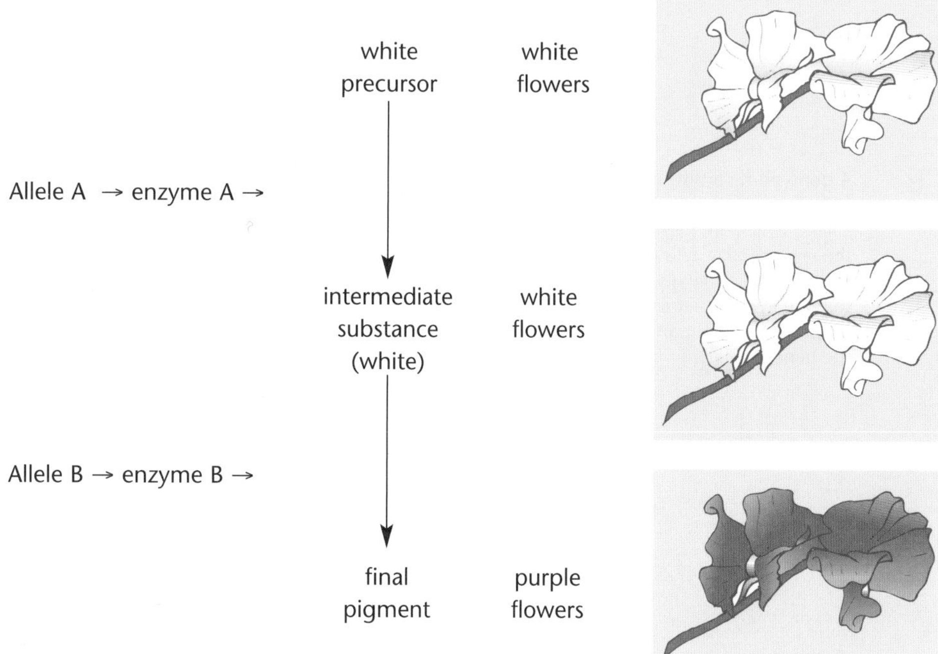

Both alleles A and B are needed to code for their respective enzymes if purple sweet pea flowers are to be produced. The alleles to consider are:

A (dominant) a (recessive) B (dominant) b (recessive)

Genotypes	aabb	aaBb	aaBB	Aabb	AaBB	AaBb	AAbb	AABb	AABB
Phenotypes	white	white	white	white	purple	purple	white	purple	purple

Without the combined effects of both A and B alleles, then the flowers are white. The reliance of one gene on another is an example of **epistasis**.

How is it possible for two white flowered plants to be crossed to give purple offspring?

Questions about epistasis usually give some data which you will need to analyse. The organisms may not be sweet pea plants but the principles remain the same.

	white		white
genotype	AAbb	×	aaBB
gametes	Ab Ab		aB aB
F_1 generation		AaBb	

100% purple flowered plants from white flowered parents.

Check out the above genotypes to find two more genotypes of white flowered plants which could be crossed to give purple offspring.

Hardy–Weinberg principle

OCR 5.1.2

The application of this principle allows us to **predict numbers of expected genotypes** in a population in the future. The principle tracks the proportion of two different alleles in the population.

Before applying the Hardy–Weinberg principle, the following criteria must be satisfied.

- There must be no immigration and no emigration.
- There must be no mutations.
- There must be no selection (natural or artificial).
- There must be true random mating.
- All genotypes must be equally fertile.

Once the above criteria are satisfied then **gene frequencies remain constant**.

A **gene pool** consists of all genes and their alleles, which are part of the reproductive cells of an organism. Only genes that are in cells that **can be passed on** are part of the gene pool.

Hardy–Weinberg principle: the terms identified

p = the frequency of the dominant allele in the population

q = the frequency of the recessive allele in the population

p^2 = the frequency of homozygous dominant individuals

q^2 = the frequency of homozygous recessive individuals

$2pq$ = the frequency of heterozygous individuals

KEY POINT

The principle is based on two equations:

(i) $p + q = 1$ (gene pool)

(ii) $p^2 + 2pq + q^2 = 1$ (total population)

Applying the Hardy–Weinberg principle

A population of *Cepaea nemoralis* (land snail) lived in a field. In a survey there were 1400 pink-shelled snails and 600 were yellow. There were two alleles for shell colour.

Y = pink shell (dominant) y = yellow shell (recessive). Snails with pink shells can be YY or Yy. Snails with yellow shells can be yy only.

phenotype	pink	yellow
genotype	YY Yy	yy

This part of the calculation is to find the frequency of the recessive and dominant alleles in the population.

$$q^2 = \frac{600}{2000}$$

$$= 0.30$$

$$q = \sqrt{0.30} = 0.55$$

But: $p + q = 1$

$$p = 1 - 0.55$$

$$= 0.45$$

So: $p^2 = 0.20$

This part of the calculation is to find the frequency of homozygous and heterozygous snails in the population.

But: $p^2 + 2pq + q^2 = 1$

$$0.20 + 0.50 + 0.30 = 1$$

YY Yy yy

Always use the $p + q = 1$ equation to calculate the frequency of alleles if you are given suitable data, e.g. 'out of 400 diploid organisms in a population there were 40 homozygous recessive individuals'. 40 organisms have 80 recessive alleles.

$$q = \frac{80}{800}$$

$$= 0.1$$

From this figure you can calculate the others.

Points to note

- These proportions can be applied to the snail populations in say, 10 years in the future.
- If there were 24 000 snails in the population, then the relative numbers would be:
  - YY $0.2 \times 24\,000 = 4800$
  - Yy $0.5 \times 24\,000 = 12\,000$
  - yy $0.3 \times 24\,000 = 7200$
- Remember that the five criteria must be satisfied if the relationship is to hold true.
- It is not possible to see which snails are homozygous dominant and which are heterozygous. They all look the same, pink! Hardy–Weinberg informs us, statistically, of those proportions.

It is also possible to apply the Hardy–Weinberg principle to a co-dominant pair of alleles. P and q are calculated by exactly the same method.

Chi-squared: a statistical test

OCR 5.1.2

When doing scientific investigations, we need to know if our results are significant or due to chance. We should not, for example, conclude that a new genetic ratio we have found represents a significant pattern for a particular cross. The χ^2 (**chi-squared**) test helps us to check out the difference between **expected** results and **actual** results. We can then state the probability that any differences between expected and actual results are due to chance or have significance.

Remember, in **co-dominance** both alleles are expressed in the phenotype.

$$\chi^2 = \sum \frac{d^2}{x}$$

d = difference between actual and expected results
x = expected results
Σ = the sum of

Consider this example

Dianthus (campion) has flowers of three different colours, red, pink and white. Two pink flowered plants were crossed and the collected seeds grown to the flowering stage.

R = red r = white (both alleles are co-dominant)

genotypes Rr × Rr

gametes R r R r

F_1 generation

	R	r
R	RR	Rr
r	Rr	rr

white 0.25
pink 0.5
red 0.25

In an examination, you may be given another term for 'actual'. It may be 'observed', but it means the same!

Numbers	RR = red flowers	Rr = pink flowers	rr = white flowers
Actual	34	84	42
Expected	40	80	40

$$\chi^2 = \frac{(40-34)^2}{40} + \frac{(80-84)^2}{80} + \frac{(40-42)^2}{40}$$
$$= 0.9 + 0.2 + 0.1$$
$$= 1.2$$

The next stage is to assess the degrees of freedom for this investigation. This value is always one less than the number of classes of results. In this case there are three classes, i.e. red, pink and white.

Degrees of freedom = (3 – 1) = 2

Now check the χ^2 value against the table.

Degrees of freedom				χ^2				
1	0.00	0.10	0.45	1.32	2.71	3.84	5.41	6.64
2	0.02	0.58	1.39	2.77	4.61	5.99	7.82	9.21
Probability that deviation is due to chance alone (significance level)	0.99 (99%)	0.75 (75%)	0.50 (50%)	0.25 (25%)	0.10 (10%)	0.05 (5%)	0.02 (2%)	0.01 (1%)

If you are given a χ^2 question in an examination you will be given a data table. A mark may be given for degrees of freedom. Remember, **10** classes of results would give **9** degrees of freedom.

What do you do with the χ^2 value?

- Go to the 0.05 level of significance (5%).
- At 2 degrees of freedom, is the χ^2 value (in this case 1.2) greater than the value given in the table (5.99)?
- If it is greater than there is a significant difference between the observed and expected data.

In Biology the 0.05 level is the one that is generally used.

Sample question and model answer

(a) Explain the difference between sex linkage and autosomal linkage. [2]

sex linkage – genes are located on a sex chromosome

autosomal linkage – genes are located on one of the other 44 chromosomes

(b) The diagram below shows part of a family tree where some of the people have haemophilia.

This type of question is a challenge! Note the key for the symbols and then apply them to the family tree. Think logically and work up and down the diagram. In your 'live' examination write on the diagram to help you work out each individual genotype asked in the question. If there are a range of possible genotypes they may be helpful.

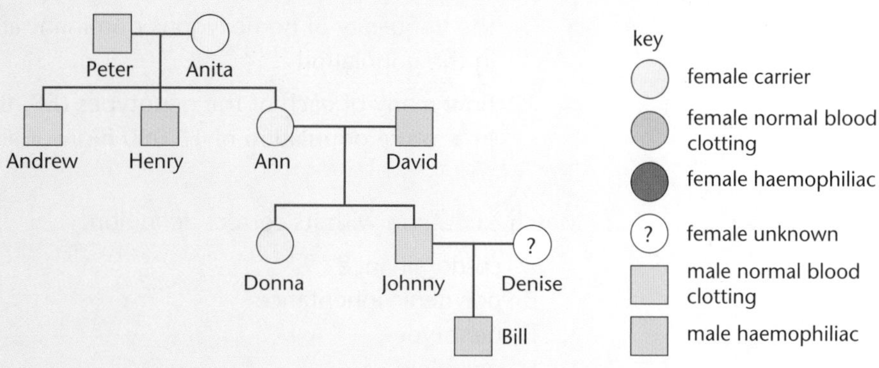

Show the possible genotypes of Denise. Give evidence from the genetic diagram to support your answer. [3]

Let H = normal blood clotting

Let h = haemophiliac trait

The genotype can be X^HX^h or X^hX^h

Reason – Johnny is X^HY so he is responsible for Bill's Y chromosome (Y chromosomes do not carry a blood clotting gene)

Working backwards, Bill is haemophiliac so Denise must have at least one X^h

She can, therefore, be X^hX^h or X^HX^h

(c) Peter and Anita had three children. Andrew was born first, then Henry and finally Ann. Use the information in the diagram to answer the questions.

(i) When could genetic counselling have been given to help Peter and Anita? [1]

After the birth of Henry.

(ii) Explain the useful information which they could have been given. [3]

Since Henry is haemophiliac his genotype is X^hY.

His father, Peter, has normal clotting blood so is X^HY and passes on a Y to his son, Henry.

His mother is not haemophiliac but must be a carrier, X^HX^h because mother passes on X^h.

We can predict 2 in 4 children will have normal clotting of blood, 1 in 4 will be female and a carrier and 1 in 4 will be haemophiliac male.

Practice examination questions

1 (a) List the criteria which must be satisfied before applying the Hardy–Weinberg principle. [4]

(b) In a population of 160 small mammals, some had a dark brown coat and the others had a light brown coat. Dark brown (B) is dominant over light brown (b). In the population there were 48 light brown individuals. Using the Hardy–Weinberg equations calculate:

(i) the frequency of homozygous dominant and heterozygous individuals in the population [3]

(ii) how many of each of the genotypes (BB, Bb, bb) there would be in a future population of 10 000 individuals. [2]

[Total: 9]

2 Match each term with its correct definition.

A co-dominance
B polygenic inheritance
C genotype
D polyploid
E somatic

(i) a cell which is not involved in reproduction [1]

(ii) a nucleus which has three or more sets of chromosomes [1]

(iii) a feature which is controlled by two or more genes, along different loci along a chromosome [1]

(iv) two alleles which are equally expressed in the organism [1]

(v) all of the genes found in a nucleus, including both dominant and recessive alleles [1]

[Total: 5]

3 The letters below represent the organic bases along the coding strand of a DNA molecule.

CCG ATT CGA TAG

(a) What term is given to each group of three bases? [1]

(b) Give **two** functions of a group of three organic bases. [2]

[Total: 3]

4 The diagram on the right shows a cell at the beginning of telophase II during meiosis.

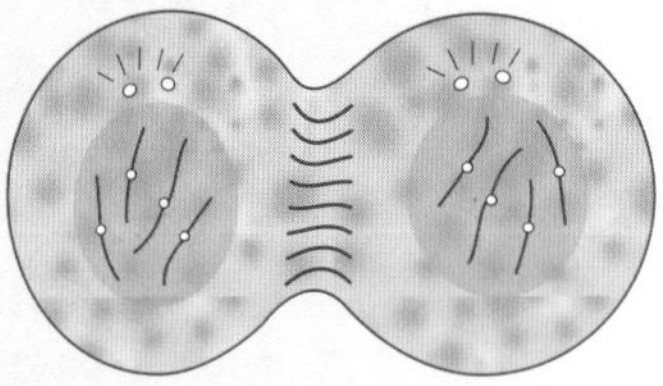

(a) How many chromosomes were there in the parent cell at the beginning of meiosis? [1]

(b) Describe **one** difference between telophase II and:

(i) telophase I of meiosis

(ii) telophase of mitosis. [2]

(c) Describe the stage immediately before telophase II. [2]

[Total: 5]

Chapter 5

Variation and selection

The following topics are covered in this chapter:

- *Variation*
- *Selection and speciation*

5.1 Variation

After studying this section you should be able to:

- *explain the different sources of variation in organisms*
- *describe different types of mutation*

LEARNING SUMMARY

Variation and mutations

OCR 5.1.1

Meiosis and sexual reproduction can produce variation in a number of ways. These include:

- segregation or independent assortment of homologous chromosomes
- chiasmata formation leading to crossing over
- random fusion of gametes.

All these processes will combine alleles in different combinations.

The environment will also contribute to variation. The combination of environmental variation and a number of genes controlling a characteristic (polygenic inheritance) will often result in a wide range of phenotypes and continuous variation. However, the only way that new alleles can be made is by mutation.

Mutation is a change in the DNA of a cell. If the cell affected by mutation is a **somatic cell**, then its effect is **restricted** to the organism itself. If, however, the mutation affects **gametes**, then the genetic change will be inherited by the future population.

Gene mutations

A gene mutation involves a change in a single gene. This is often a point mutation.

Bases can change along DNA and this may cause mutation. One changed base along the coding strand of DNA may have a sequential effect of changing most amino acids along a polypeptide.

before mutation
TTA CCG GCC ATC

after mutation
ATT ACC GGC CAT C

This is addition!

DNA codes for the sequence of amino acids along polypeptides and ultimately the characteristics of an organism. Each amino acid is coded for by a triplet of bases along the coding strand of DNA, e.g. TTA codes for threonine. The change in a triplet base code can result in a new amino acid, e.g. ATT codes for serine. This type of DNA change along a chromosome is known as a **point mutation**. A point mutation involves a change in a single base along a chromosome by **addition** (insertion), **deletion** or **inversion**.

KEY POINT

An example of a gene mutation is a change in the DNA coding for the protein haemoglobin. This can cause sickle-cell anaemia.

Sometimes a point mutation may not cause a change in the phenotype. This is because the genetic code is degenerate. Often one amino acid has more than one triplet coding for it. Therefore a change in a base may not change the amino acid.

Key points from AS

- **Variation**
  Revise AS pages 83–84

More mutations are shown below. Each section of DNA along the chromosomes is shown by organic bases.

Addition

before
TTA CCG GCC ATC

after
CCG TTA CCG GCC ATC

A new triplet has been added. If a triplet is repeated it is also duplication.

Deletion

before
TTA CCG GCC ATC

after
TTA CCG GCC

Inversion

before
TTA CCG GCC ATC

after
TTA CCG GCC **CTA**

CTA codes for a new amino acid.

Translocation

before
TTA CCG GCC ATC

after
TTA CCG GCC ATC **CAT**

CAT broke away from another chromosome.

Chromosome mutations

If a complete chromosome is added or deleted, this is a **chromosomal mutation.** Sometimes something goes wrong during meiosis and both members of a homologous pair of chromosomes move to the same pole. This produces a gamete with an extra chromosome and, after fertilisation, the zygote has an extra chromosome. This is called aneuploidy. A example is Down's syndrome where a person has an additional chromosome, totalling 47 in each nucleus rather than the usual 46.

If the spindle fails altogether, then an individual can be produced with whole extra sets of chromosomes. This is called polyploidy and is important in plant evolution.

What causes mutations?

All organisms tend to mutate randomly, so different sections of DNA can appear to alter by chance. The appearance of such a random mutation is usually very rare, typically one mutation in many thousands of individuals in a population. The rate can be increased by **mutagens** such as:

- **Ionising radiation** – including ultra violet light, X-rays and α(alpha), β(beta) and γ (gamma) rays and neutrons. These forms of radiation tend to dislodge the electrons of atoms and so disrupt the bonding of the DNA which may re-bond in different combinations.
- **Chemicals** – including asbestos, tobacco, nitrous oxide, mustard gas and many substances used in industrial processes such as vinyl chloride. Many pesticides are suspected mutagens. Dichlorvos, an insecticide, is a proven mutagen. Additionally, colchicine is a chemical derived from the Autumn crocus, *Colchicium*, which stimulates the development of extra sets of chromosomes.

Are mutations harmful or helpful?

An individual mutation may be either harmful or helpful. When tobacco is smoked, it can increase the rate of mutation in some somatic cells. The DNA disruption can result in the formation of a cell which divides uncontrollably and causes the disruption of normal body processes. This is **cancer**, and can be lethal. The presence of certain genes called **oncogenes** is thought to increase the rate of cell division and lead to cancer.

Chrysanthemum plants have a high rate of mutation. A chrysanthemum grower will often see a new colour flower on a plant, e.g. a plant with red flowers could develop a side shoot which has a different colour, such as bronze. Most modern chrysanthemums appeared in this way, production being by asexual techniques.

Some mutated human genes have, through evolution, been successful. Many successful mutations contributed to the size of the cerebrum which is proportionally greater in humans than in other primates.

5.2 Selection and speciation

After studying this section you should be able to:

- *understand the process of natural selection*
- *predict population changes in terms of selective pressures*
- *understand a range of isolating mechanisms and how a new species can be formed*
- *understand the difference between allopatric and sympatric speciation*
- *explain the difference between natural selection and artificial selection*

LEARNING SUMMARY

Natural selection

OCR 5.1.2

Throughout the biosphere, communities of organisms interact in a range of ecosystems. Darwin travelled across the world in his ship, the *Beagle*, observing organisms in their habitats. In 1858 Darwin published *On the Origin of Species*. In this book he gave his theory of **natural selection**.

The key features of this theory are that as organisms interact with their environment:

- individual organisms of populations are not identical, and can **vary in both genotypes and phenotypes**
- **some organisms survive** in their environment other organisms **die** before reproducing, effectively being **deleted from the gene pool**
- surviving organisms **go on to breed** and **pass on their genes** to their offspring
- this **increases the frequency of the advantageous genes** in the population.

Learn this theory carefully then apply it to the scenarios given in your examination. Candidates often identify that some organisms die and others survive, but few go on to predict the inheritance of advantageous genes and the consequence to the species.

Consider these factors

- Adverse conditions in the environment could make a species extinct, but a range of genotypes increases the chances of the species surviving.
- Different genotypes may be suited to a changing environment, say, as a result of global warming.
- A variant of different genotype, previously low in numbers, may thrive in a changed environment and increase in numbers.
- Where organisms are well suited to their environment they have adaptations which give this advantage.
- If other organisms have been selected against, then more resources are available for survivors.
- Breeding usually produces many more offspring than the mere replacement of parents.
- Resources are limited so that competition for food, shelter and breeding areas takes place. Only the fittest survive!

What is selective pressure?

Selective pressure is the term given to a factor which has a direct effect on the numbers of individuals in a population of organisms, for example:

'It is late summer and the days without rainfall have caused the grassland to be parched. There is little food this year.'

In this example, the fact that the numbers of herbivores decrease is ***another*** selective pressure. This time numbers of predators may decrease.

Here the **selective pressure** is a **lack of food** for the herbivores. Species which are **best adapted** to this habitat **compete** well for the limited resources and go on to survive. Within a species there is a further application of the selective pressure as weaker organisms perish and the strongest survive.

Mutations are random

Considering these examples, it is no wonder that candidates seem to consider that the organisms actively adapt to develop in these ways. They suggest that the organisms themselves have control to make active changes. **This is not so! There is no control, no active adaptation.**

New genes can appear in a species for the first time, due to a form of mutation. Over thousands of years, repeated natural selection takes place, resulting in superb adaptations to the environment.

- The Venus fly trap with its intricate leaf structures captures insects. The insects decompose, supplying minerals to the mineral deficient soil.
- Crown Imperial lilies (*Fritillaria*) produce colourful flowers, and a scent of stinking, decomposing flesh. Flies are attracted and help pollination.
- The bee orchid flower is so like a queen bee that a male will attempt mating.

New genes appear by CHANCE!

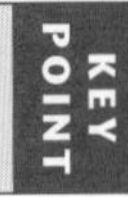

Selective pressures and populations

To find out more about the effects that selective pressures can have, the **normal distribution** must be considered. The distribution below is illustrated with an example.

The mean value is at the peak. There are fewer tall and short individuals in this example. A taller plant intercepts light better than a shorter one.

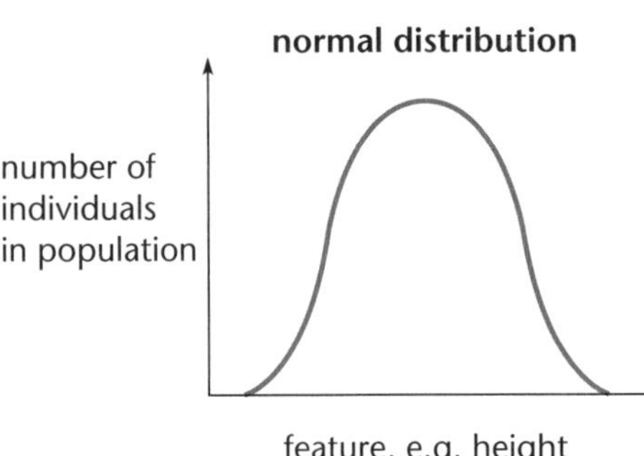

The further distributions below show effects of selective pressures (shown by the blue arrows). Each is illustrated with an example.

Selective pressure at both ends of the distribution causes the extreme genotypes to die. This maintains the distribution around the mean value. Mean wing length is better for flight, better for prey capture.

Selective pressure results in death of slower animals. Many die out due to predators. Faster ones (with longer legs) pass on advantageous genes. Distribution moves to the right as the average individual is now faster.

Selective pressure results in the death of organisms around the mean value. In time this can lead to two distributions. Long fur is adapted to a cold temperature and short fur to a warm temperature. The mean is suited to neither extreme.

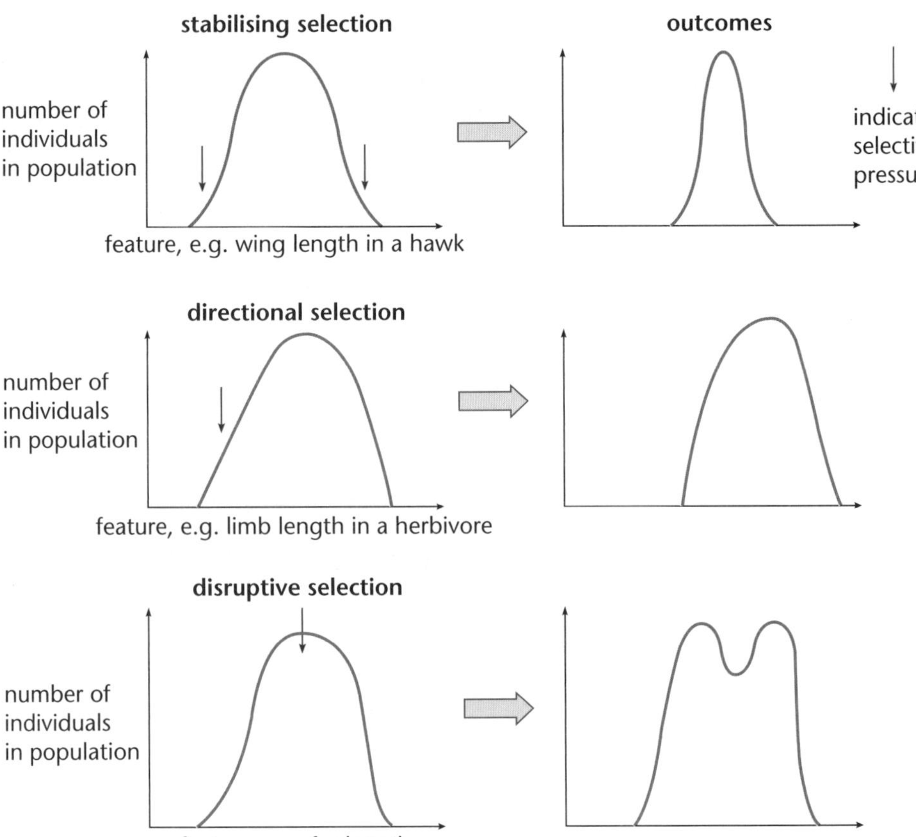

Speciation

OCR 5.1.2

The previous example of disruptive selection showed how two extreme genotypes can be selected. Continued selection against individuals around the former mean genotype finally results in two discrete distributions. This division into two groups may be followed by, for example, advantageous mutations. There is a probability that, in time, the two groups will become incompatible and unable to breed successfully. They have become a new species. The **development of new species is called speciation**.

To enable enough genetic differences to build up between the two groups, they must be isolated to stop them breeding. This can happen in a number of ways.

Geographical isolation

This will help you. Different finches evolved on different islands, but they did have a common ancestor.

This takes place when a population becomes divided as a result of a physical barrier appearing. For example, a land mass may become divided by a natural disaster like an earthquake or a rise in sea level. Geographical isolation followed by mutations can result in the formation of new species. This can be illustrated with the finches of the Galapagos islands. There are many different species in the Galapagos islands, ultimately from a common ancestral species. Clearly new species do form after many years of geographical isolation. This is **allopatric speciation**.

Reproductive isolation

A new pheromone is produced by several antelopes as a result of a mutation. The mainstream individuals refuse to mate as a result of this scent. An isolated few do mate. This is reproductive isolation.

This is a type of genetic isolation. Here the formation of a new species can take place in the same geographical area, e.g. mutation(s) may result in reproductive incompatibility. A new gene producing, for example, a hormone, may lead an animal to be rejected from the mainstream group, but breeding may be possible within its own group of variants. The production of a new species by this mechanism is known as **sympatric speciation**.

Artificial selection

OCR 5.1.2

In natural selection, the selection pressure comes from the organism's environment. In **artificial selection**, humans choose which organisms are allowed to reproduce. This is **selective** breeding to improve specific domesticated animals and crop plants.

Important points are:

Artificial selection is not the only way to improve animals and plants. Genetic modification is another method. A variety of soya bean plants now has resistance to selective herbicides.

- people are the **selective agents** and choose the parent organisms which will breed
- the organisms are chosen because they have **desired characteristics**
- the aim is to incorporate the desired characteristics from both organisms in their offspring
- the offspring must be **assessed** to find out if they have the desired combination of improvements (there is **no guarantee** that a cross will be successful!)
- offspring which have suitable improvements are used for breeding, the others are deleted from the gene pool (not allowed to breed).

Most modern crops have been produced by artificial selection. Modern wheat is one example. The Brussels sprout variety opposite was produced in this way. Many trials were carried out before the new variety was offered for sale.

Can you suggest four excellent features offered by this new variety?

Brilliant NEW FOR 2001
F1 Hybrid A brand new early cropping variety which produces dense, dark green buttons of excellent quality in September and October. Suitable for a wide range of soil types it also has a high resistance to powdery mildew and ring spot. Good for freezing. **2152 pkt £2.10**

All modern racehorses have been artificially selected. Champion thoroughbred horses are selected for breeding on the basis of success in races. Only the best racehorses are actually entered in races. The fastest horses, at various distances, win races and the right to breed. Continual improvement results as the gene pool is consistently strengthened. Modern dairy cows have also been produced by artificial selection.

Sample question and model answer

The graphs below show the height of two pure breeding varieties of pea plant, Sutton First and Cava Late.

Continuous variation can confuse you sometimes when examiners display the data in categories as histograms. **This is not discontinuous!**

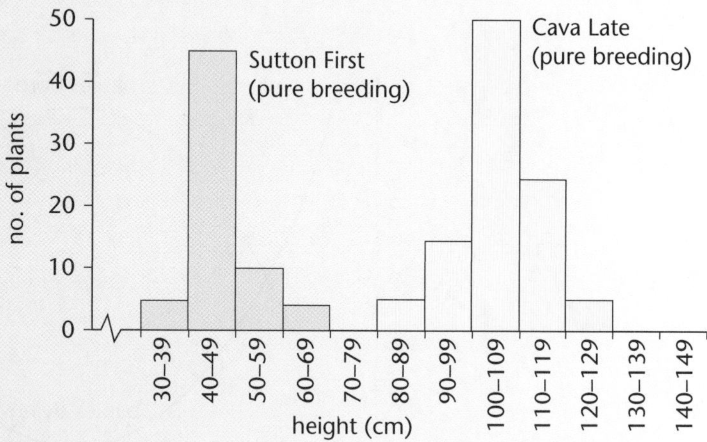

(a) (i) Which types of variation are shown by the pea variety, Sutton First? Give evidence from the bar graph to support your answer. [4]

Continuous variation – this is shown by the increase across the distribution (even though the peas are pure breeding).

Environmental variation – shown by the range of different heights.

(ii) Which type of variation is shown **between** varieties Sutton First and Cava Late? Give evidence from the bar graphs to support your answer. [2]

Discontinuous variation – the two distributions are separate and do not intersect.

(iii) Both Sutton First and Cava Late have compatible pollen for cross-breeding. Suggest why they do **not** cross breed. [1]

As implied by the names, Sutton First flowers before Cava Late, so that the flowers are not ready at the same time.

When you are asked to 'suggest', then a range of different plausible answers are usually acceptable.

(b) Plant geneticists considered that many years ago the two varieties of pea had the same ancestor.

(i) Suggest what, in the ancestor, resulted in the difference in height of the two varieties? [1]

mutation

(ii) Suggest what caused this change. [1]

radiation/random processes

(c) (i) Define polygenic inheritance. [1]

The inheritance of a feature controlled by a number of genes (not just a gene at one locus).

(ii) Which type of variation is a consequence of polygenic inheritance? [1]

continuous variation

Practice examination questions

1 The graph shows the birth weight of babies born in a London hospital between 1935 and 1946. It also shows the chance of the babies dying within two months of birth.

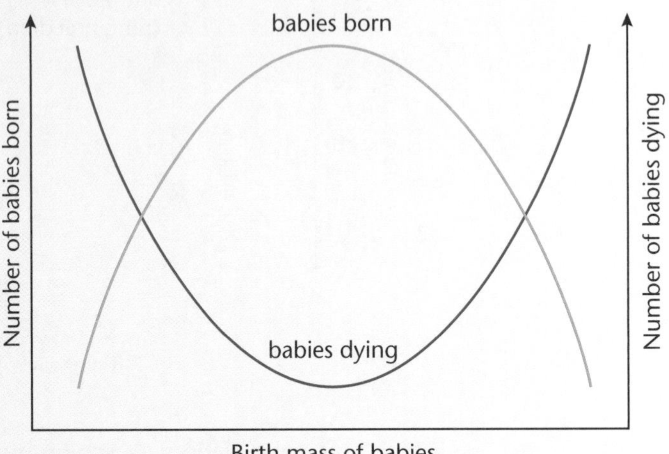

(a) What type of variation is shown by the birth weight of the babies? [1]

(b) What factors decide the birth weight of a baby? [2]

(c) Scientists argue that the information in the graphs shows that stabilising selection acts on the birth weight of babies. Explain why they think this. [3]

(d) The data was collected between 1935 and 1946. Modern medical techniques may have altered the selection pressure on the birth weight of babies. Explain why. [2]

[Total: 8]

2 (a) Explain the difference between allopatric and sympatric speciation. In each instance use an example to illustrate your answer. [6]

(b) How is it possible to find out if two female animals are from the same species? [2]

[Total: 8]

Chapter 6

Biotechnology and genes

The following topics are covered in this chapter:

- Growing microorganisms
- Mapping and manipulating genes
- Cloning

6.1 Growing microorganisms

After studying this section you should be able to:

- *describe how microorganisms are grown in ideal conditions in fermenters*
- *describe the main features of batch and continuous culture*

LEARNING SUMMARY

Microorganisms and fermenters

OCR 5.2.2

Before describing how substances are commercially produced it is necessary to consider the meaning of the term, **biotechnology**.

> Biotechnology is the use of organisms and biological processes to supply nutrients, other substances and services to meet human needs. **Fermentation** is a key process in biotechnology using microorganisms to produce traditional products such as ethanol and more recently substances such as pharmaceutical chemicals and the enzymes for biological washing powder.
>
> KEY POINT

Modern industrial fermenters

- There are several different types of fermenter used to grow microorganisms on a large scale. They all have the common purpose of producing food, or chemicals such as antibiotics, hormones, or enzymes. The fermenter below shows a typical design.

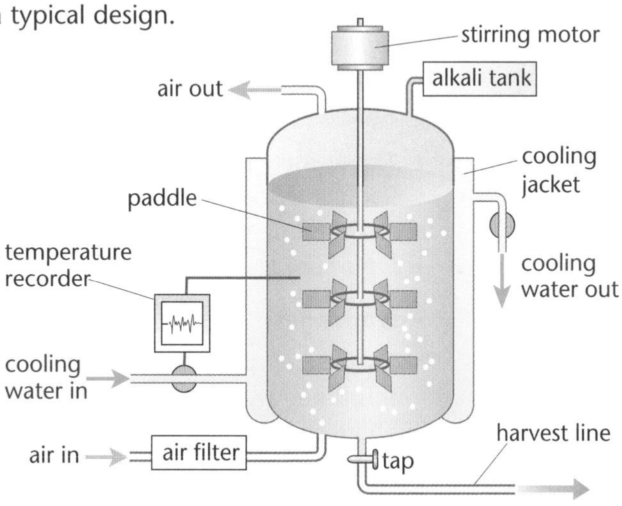

New transgenic organisms are continually being developed. Many hit the headlines in the media. Be aware that the examiners may use these high-profile organisms in questions. Do not become disorientated – the principle is always the same.

Conditions inside fermenters should be suitable for the optimal metabolism and rapid reproduction of the microorganisms. Products should be harvested without contamination. Note the conditions which need to be controlled.

- Fermenters are sterilised using steam before adding nutrients and the microorganisms used during the process. Conditions are **aseptic**.
- Nutrients which are specifically suited to the needs of the microorganisms are supplied.
- Air is supplied if the process is aerobic. This must be filtered to avoid

contamination from other microorganisms.

- Temperature must be regulated to keep the microorganisms' enzymes within a suitable range. An active 'cooling jacket' and heater, both controlled via a thermostat, enable this to be achieved.
- pH must remain close to the optimum. Often the development of low pH during fermentation would result in the process slowing down or stopping. The addition of alkaline substances allows the process to continue and maximises yield.
- Paddle wheel mixing or 'bubble agitation' make sure that the microorganisms meet the required concentrations of nutrients and oxygen.

This graph shows the rate of production of a primary metabolite. These are produced by a microbe as part of its normal growth. Secondary metabolites such as antibiotics are usually produced after the main growth phase.

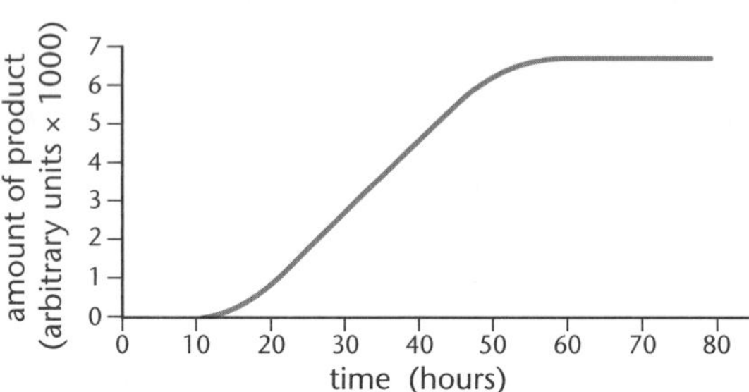

Batch culture

Batch culture takes place in a **closed** vessel. The microorganism is allowed to grow and then stopped and the product is removed.

Advantages

- If the culture becomes contaminated in any way, just one batch is spoiled.
- The fermenter can be used for a variety of fermentation processes, e.g. different antibiotics.

Disadvantages

- At the end of every production period, **shut down** takes place. The vessel needs to be cleaned and re-sterilised. This lost time can be expensive to the company.
- Often the product, waste substances and unused nutrients are mixed together, e.g. in penicillin production. Product removal is made more difficult by these contaminants.

Continuous culture

Continuous culture takes place in an open fermenter.

- The fermenter is steam sterilised.
- Regular amounts of sterile nutrients are added.
- At the same time regular amounts of product are removed.
- Optimum levels of pH, oxygen, nutrients and temperature are maintained.

Advantages

- The rate of growth of the microbial population is kept at a maximum level: this is known as the **exponential rate**.
- There is **no** regular pattern of **shut down**.

Disadvantages

- Maintaining the levels at **optimal levels** is difficult.
- **Regular** sampling is necessary for quality control, ensuring that chemicals are in equilibrium, and contaminants are absent.
- There is more chance of contaminants entering due to regular input and output.

Immobilised enzymes

After completion of an enzyme catalysed reaction the enzyme remains unchanged and can be used again. Unfortunately the enzyme can contaminate the product, as it can be difficult to separate out from the reaction mixture. For this reason

immobilised enzymes have been developed. They are used as follows:

- enzymes are attached to insoluble substances such as resins and alginates
- these substances usually form membranes or beads and the enzymes bind to the outside
- substrate molecules readily bind with the active sites and the normal reactions go ahead
- the immobilised enzymes are easy to recover, remaining in the membranes or beads
- there is no contamination of the product by free enzymes
- expensive enzymes are re-used
- processes can be continuous unlike batch, where the process is stopped for 'harvesting'.

Progress check

State **two** differences between continuous and batch production.

- Batch production is shut down on a more regular basis.
- In continuous production, nutrients are added on a regular basis, and products are removed in similar quantities. In batch production product retrieval is at the end rather than during the process.

6.2 Cloning

After studying this section you should be able to:

- *describe how plants can be cloned*
- *explain why it is harder to clone animals*
- *describe the possible uses of cloning*

LEARNING SUMMARY

Producing clones

OCR 5.2.1

As discussed on page 63 the cells in the early embryo are called **totipotent**. This means that they can develop into any type of cell. However, as the embryo develops, cells become specialised and lose this ability. In mature plants, many cells remain totipotent. This means that it is quite easy to produce new plants from sections of plant tissue. The plants that are produced are called clones. In mature animals these totipotent or **stem cells** are harder to find.

Clones are genetically identical to their parent. Plants can clone themselves naturally or can be cloned artificially by processes such as **tissue culture** or **micropropagation.**

KEY POINT

Advantages of micropropagation

- Generating new plants from the apical meristem tissue eliminates many plant viruses, so usually, virus-free plants are produced.
- If the material is available the process can take place at any time of the year.
- Even a tiny explant or callus can be cut into pieces and sub-cultured.

The diagrams below outline the process.

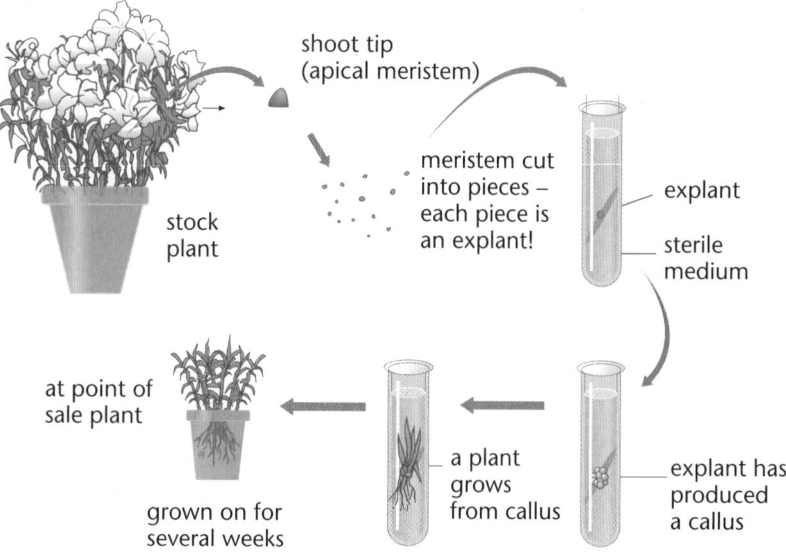

Cloning animals

Since Dolly was produced, many other types of mammals have been cloned including monkeys, dogs and cats.

Cloning animals is much more difficult because of the problem of finding suitable stem cells. There are two possibilities:

- Take an early embryo and split up the ball of cells. At this early stage the cells will be able to develop into separate cloned individuals.
- It is now possible to clone many animals from adult body cells. This is performed by nuclear transfer and was first used to produce Dolly the sheep in 1996. The diagram shows how it is done.

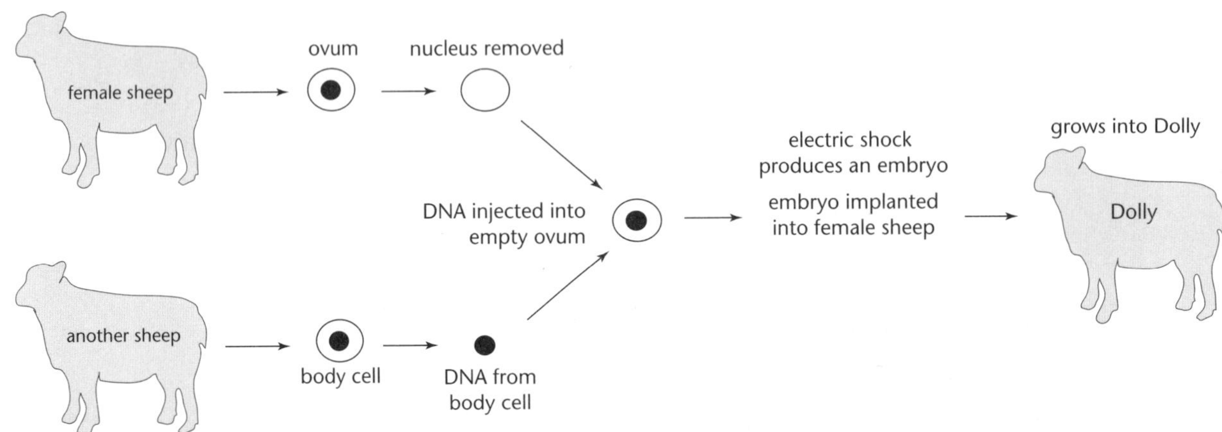

Uses of cloning

Cloning animals by splitting up an embryo may have limited uses. As the embryo has been produced by sexual reproduction, the genetic make-up of the clones is uncertain. Cloning animals from adult body cells may be much more useful. There are two main possibilities:

- **Reproductive cloning** This could be used to produce identical copies of endangered animals, animals with desired characteristics or even embryos for infertile human couples.
- **Therapeutic cloning** This may be used to provide a source of stem cells from the early embryo. The stem cells may be used to treat degenerative diseases. The embryo is then destroyed.

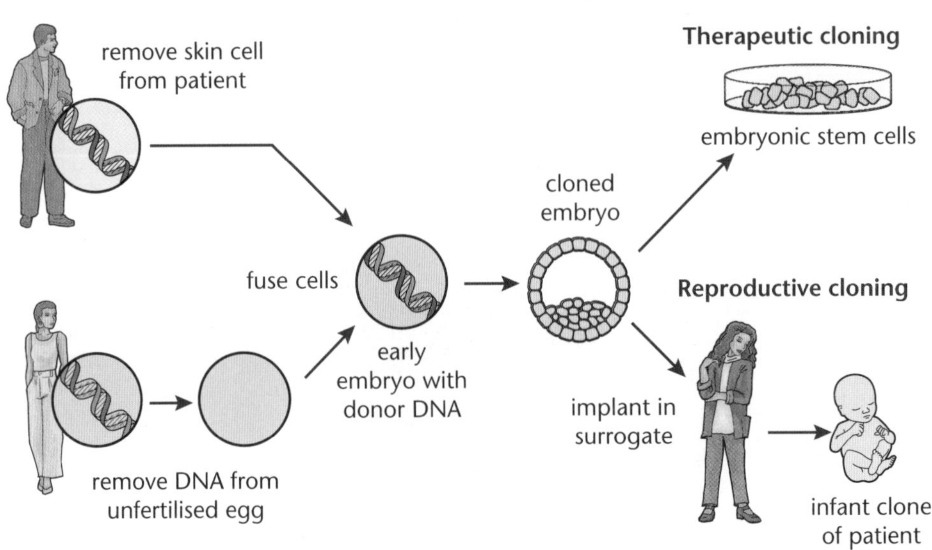

6.3 Mapping and manipulating genes

After studying this section you should be able to:

- *describe a range of techniques used in mapping genes*
- *describe how genes can be transferred between a range of different organisms*

LEARNING SUMMARY

Identifying genes

OCR 5.2.3

In order to identify the genes of an individual, a number of different processes are used.

Polymerase chain reaction

The polymerase chain reaction (PCR) is used to make numerous copies of a section of DNA. This is called **amplifying** the DNA. It uses the principle of semiconservative replication of DNA to produce new molecules that can in turn act as templates to produce more molecules. This therefore sets up a chain reaction. The enzyme DNA polymerase is used to copy the DNA.

Electrophoresis

This is used to separate sections of DNA according to their size.

Enzymes called restriction endonucleases can be used to cut up an organism's DNA.

- DNA sections are put into a well in a slab of agar gel.
- The gel and DNA are covered with buffer solution which conducts electricity.
- Electrodes apply an electrical field.
- Phosphate groups on DNA are negatively charged causing DNA to move towards the anode.
- Smaller pieces of DNA move more quickly down the agar track; larger ones move more slowly, leading to the formation of bands.

Genetic fingerprinting

PCR and electrophoresis have many applications. DNA is highly specific so the bands produced using this process can help with identification. In some crimes, DNA is left at the scene. Blood and semen both contain DNA specific to an individual. DNA evidence can be checked against samples from suspects. This is known as **genetic fingerprinting**. Genetic fingerprinting can be used in paternity disputes. Each band of the DNA of the child must correspond with a band from *either* the father or the mother.

Isolating genes

Along chromosomes are large numbers of genes. Scientists may need to identify and isolate a useful gene; one way of doing this is to use the enzyme **reverse transcriptase**. This is produced by viruses known as **retroviruses**. Reverse transcriptase has the ability to help make DNA from mRNA.

Stage 1

When a polypeptide is about to be made at a **mRNA** ribosome, reverse transcriptase allows a strand of its coding DNA to be made.

Stage 2

The single stranded DNA is parted from the mRNA.

Stage 3

The other strand of DNA is assembled using DNA polymerase.

Using this principle, the exact piece of DNA which codes for the production of a vital protein can be made.

Progress check

1 A length of DNA was prepared and then electrophoresis was used to separate the sections. The statements below describe the process of electrophoresis but they are in the wrong order. Write the letters in the correct sequence.

A electrodes apply an electrical field
B DNA sections are put into a well in a slab of agar gel
C smaller pieces of DNA move more quickly down the agar track with larger ones further behind
D the gel and DNA are then covered with buffer solution which conducts electricity
E restriction endonucleases can be used to cut up the DNA

2 Reverse transcriptase is an enzyme which enables the production of DNA from RNA. Work out the sequence of organic bases along the DNA of the following RNA sequence.

A A U GCCCGGAUU

1 E B D A C
2 RNA AAUGCCCGGAUU DNA$_2$ AATGCCCGGATT DNA$_1$ TTACGGGCCTAA

The Human Genome Project

The Human Genome Project is an analysis of the complete human genetic make-up, which has mapped the organic base sequences of the nucleotides along our DNA.

A brief history

- 1977 Sanger devised DNA base sequencing.
- 1986 The Human Genome Project was initiated in the USA and the UK.
- 1996 30 000 genes were mapped.
- 1999 one billion bases were mapped including all of chromosome 22.
- 2000 chromosome 21 was mapped with the human genome almost complete.
- 2001 human genome mapping complete.

Some important points

- The genome project will sequence the complete set of over 100 000 genes.
- Only around 5% of the base pairs along the DNA actually result in the expression of characteristics. These DNA sequences are known as **exons**.
- 95% of DNA base sequences are not transcribed and do not appear to be involved in the expression of characteristics. These are known as **introns**.
- Introns do not outwardly seem to be responsible for characteristics. It is likely that they may be regulatory, perhaps in multiple gene role.

Effects of single nucleotide polymorphism

Example

5 base sequences from five people →

GTATAGCCGCAT 1
GTATAGCCGCAT 1
GTATAGCCGCAT 1
GTATAGCCGCCT 2
GTATAGCCGCCT 2

Version 1 = ●
Version 2 = ●

Proportion of the SNP in healthy members of population:

Proportion of the SNP in diseased members of population:

A greater incidence of an SNP in people with a disease may point to a cause.

Single nucleotide polymorphisms (SNPs)

Around 99.9% of human DNA is the same in all individuals. Merely 0.1% is different! The different sequences in individuals can be the result of **single nucleotide polymorphism**. One base difference from one individual to another at a site may have no difference. Up to a maximum of six different codons can code for one amino acid. An SNP will not necessarily have any effect.

Some SNPs do change a protein significantly. Such changes may result in genetic disease, resistance or susceptibility to disease.

How can the mapping of SNPs be useful?

- The mapping of SNPs along chromosomes signpost where base differences exist.
- Across the gene pool a pattern of SNP positions will be evident.
- There may be a high frequency of common SNPs found in the DNA of people with a specific disease.
- This highlights interesting sites for future research and will help to find answers to genetic problems.

Benefits obtained from the Human Genome Project

Ultimately, the human genome data will be instrumental in the development of drugs to treat genetic disease. Additionally, by analysis of parental DNA, it will be possible to give the probability of the development of a specific disease or susceptibility to it, in offspring. Fetal DNA, obtained through amniocentesis or by chorionic villi sampling, will give genetic information about an individual child.

Genetic counsellors will have more information about an individual than ever before. Companies will be able to produce 'designer drugs' to alleviate the problems which originate in our DNA molecules. Soon the race will begin to produce the first crop of drugs to treat or even cure serious genetic diseases. Look to the media for progress updates.

Manipulating DNA

OCR 5.2.3

Scientists have developed methods of manipulating DNA. It can be transferred from one organism to another. Organisms which receive the DNA then have the ability to produce a new protein. This is one example of **genetic engineering**.

> **KEY POINT**
>
> The genetic code is universal. This means that it is possible to move genes from one organism to another and the recipient organism may be from a different species. The DNA will still code for the same protein.

Genes have now been transferred to and from many different types of organisms.

Here are some examples:

- From humans to bacteria: this technique produced the first commercially available genetically engineered product, insulin.
- From plants to plants: this technique has been used to produce GM crops such as Golden Rice that contain vitamin A.
- Into humans: this technique may be successful in treating genetic conditions such as cystic fibrosis (but it is not a cure).

Gene transfer to bacteria

The gene which produces human insulin was transferred from a human cell to a bacterium. The new microorganism is known as a **transgenic bacterium**. The process which follows shows how a human gene can be inserted into a bacterium.

The human gene for insulin is produced using an enzyme called **reverse transcriptase**. This converts the mRNA coding for insulin back into DNA. In this way all the introns are removed. Then the DNA can be inserted.

Restriction endonucleases are produced by some bacteria as a defence mechanism. They cut up the DNA of invading viruses. This can be exploited during gene transfer.

Note that **both** the donor DNA and recipient plasmid DNA are cut with the same enzyme. This allows the new gene to be a matching fit.

1 An enzyme known as **restriction endonuclease** cuts the DNA and the gene was removed. Each time a cut was made the two ends produced were known as 'sticky ends'.

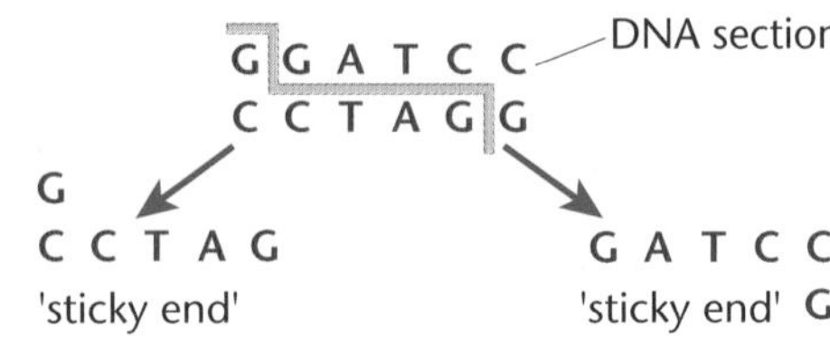

2 Circles of DNA called **plasmids** are found in bacteria.

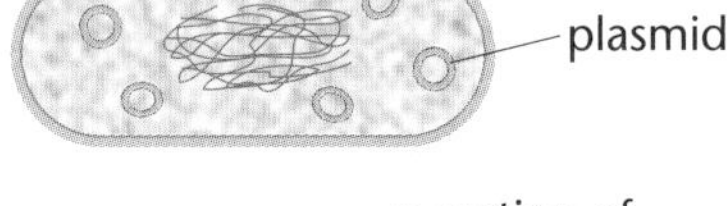

3 A plasmid was taken from a bacterium and cut with the same restriction endonuclease.

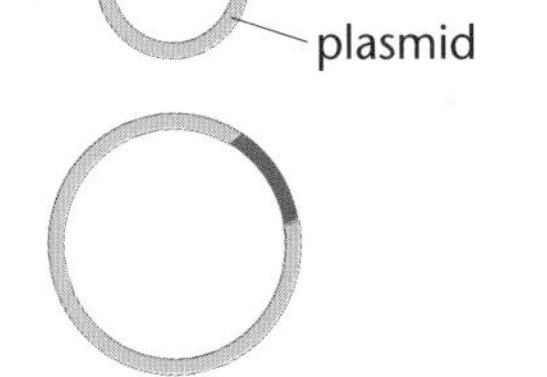

4 The human gene was inserted into the plasmid. It was made to fix into the open plasmid by another enzyme known as **ligase**.

Many exam candidates fail to state that the plasmids are cloned inside the bacterium.

5 The plasmid **replicated** inside the bacterium.

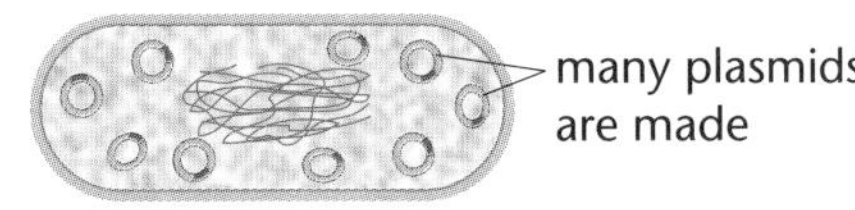

The bacteria themselves are also cloned. There may be two marks in a question for each cloning point!

6 Large numbers of the new bacteria were produced. Each was able to secrete perfect human insulin, helping diabetics all over the world.

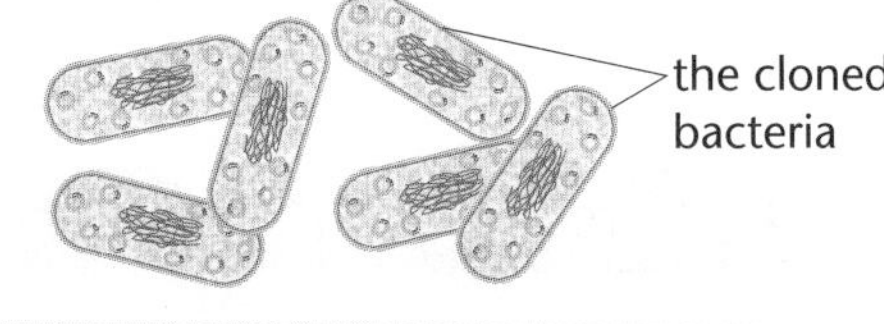

Gene transfer to plants

Inserting genes into crop plants is becoming increasingly important in meeting the needs of a rising world population. One example of this is the production of a type of rice called Golden Rice. This contains a gene that produces vitamin A. The aim is to prevent vitamin A deficiency which can lead to blindness.

In **plants** there is an important technique which uses a **vector** to insert a novel gene. The vector is the bacterium *Agrobacterium tumefaciens*.

KEY POINT

Agrobacterium tumefaciens

- This is a **pathogenic bacterium** which **invades** plants forming a gall (abnormal growth).
- The bacterium contains **plasmids** (circles of DNA) which carry a gene that stimulates tumour formation in the plants it attacks.
- The part of the plasmid which does this is known as the **T-DNA region** and can insert into any of the chromosomes of a host plant cell.
- Part of the T-DNA controls the production of two growth hormones, auxin and cytokinin.
- The extra quantities of these hormones stimulate rapid cell division, the cause of the tumour.

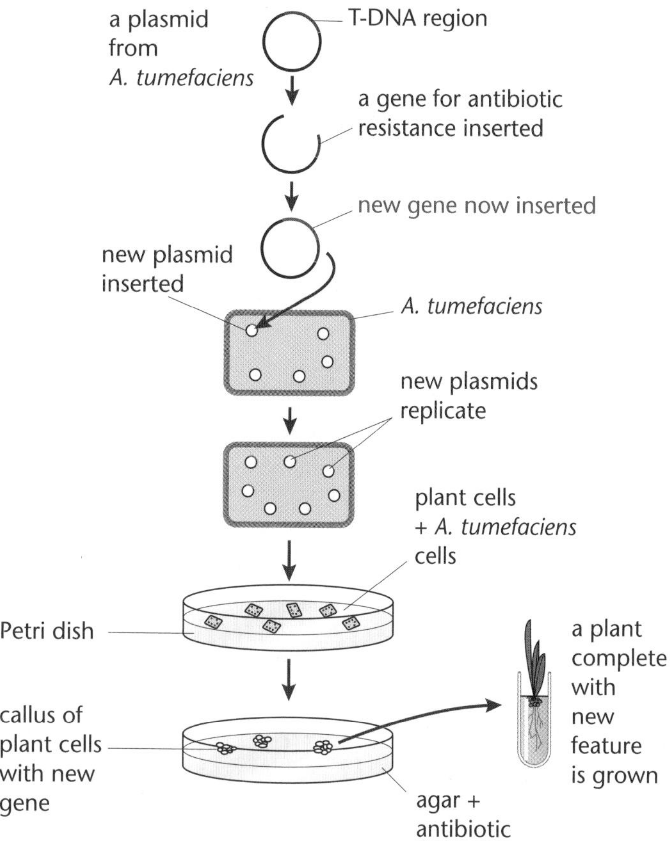

How can *Agrobacterium tumefaciens* be used in gene transfer?

KEY POINT

- Firstly, the DNA section controlling auxin and cytokinin was deleted, tumours were not formed, and cells of the plant retained their normal characteristics.
- A gene which gave the bacterial cell **resistance to a specific antibiotic** was inserted into the T-DNA position.
- The **useful gene** (e.g. the gene for vitamin A) was **inserted into a plasmid.**
- Plant cells, minus cell walls, were removed and put into a Petri dish with nutrients and *A. tumefaciens*, which contained the engineered plasmids.
- The cells were **incubated** for several days, then transferred to another Petri dish containing nutrients plus the specific antibiotic.
- **Only plant cells with antibiotic resistance and the desired gene grew.**
- Any surviving cells grew into a callus, from which an adult plant formed, complete with the transferred gene.

The principle of using *A. tumefaciens* can be used in gene transfer to many different plants. Applications are at an early stage of development.

Inserting genes into humans

The idea of changing a person's genes in order to cure genetic disease is called **gene therapy**.

There are two main possibilities:

- **Somatic cell therapy** in which the genes are inserted into the cells of the adult where they are needed.
- **Germ line gene therapy** involves changing the genes of the gametes or early embryo. This means that all the cells of the organism will contain the new gene.

Genes can also be inserted into animals to prevent their organs being rejected if used for transplants into people. This is called **xenotransplantation**.

Sample question and model answer

Look out for transgenic stories in the media. The principles are often the same. This will prepare you for potentially new ideas in your 'live' examinations. You could encounter the same account!

The diagrams below show the transfer of a useful gene from a donor plant cell to the production of a transgenic crop plant. The numbers on the diagram show the stages in the process.

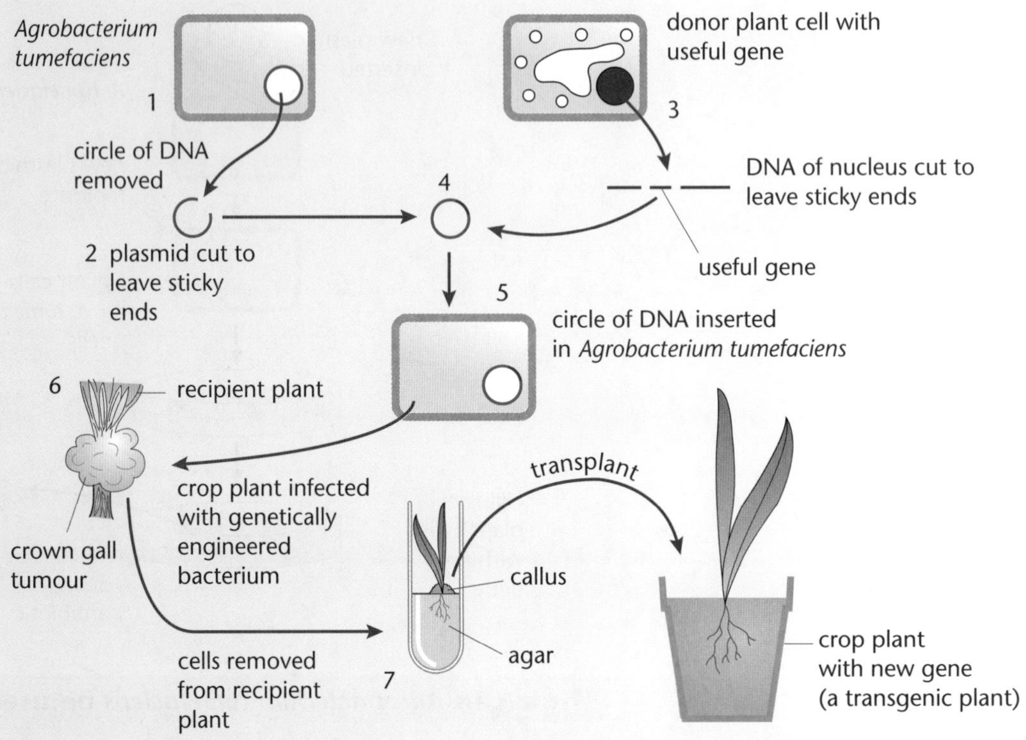

(a) Give the correct name for the circle of DNA found in the bacterium, *A. tumefaciens*. [1]

plasmid

(b) The same enzyme was used to cut the DNA of the bacterium and of the plant cell.

(i) Name the type of enzyme used to cut the DNA. [1]

restriction endonuclease

(ii) Explain why it is important to use exactly the same enzyme at this stage. [2]

The same enzyme produces the same sticky ends.

Complementary sticky ends on the donor gene bind with the sticky ends of the plasmid.

This question covers key techniques in gene transfer. Be prepared for your examination.

(iii) Which type of enzyme would be used to splice the new gene into the circle of DNA? [1]

ligase

(c) How was the new gene incorporated into the DNA of the crop plant cells? [2]

Crop plant infected by genetically engineered bacterium.

The DNA of bacterium causes a change in the DNA of the crop plant to produce the gall or tumour cells.

(d) How would you know if the gene had been transferred successfully? [1]

The feature would be expressed in the transgenic plants.

Practice examination questions

1 The diagram below shows an industrial fermenter used to produce the antibiotic, penicillin.

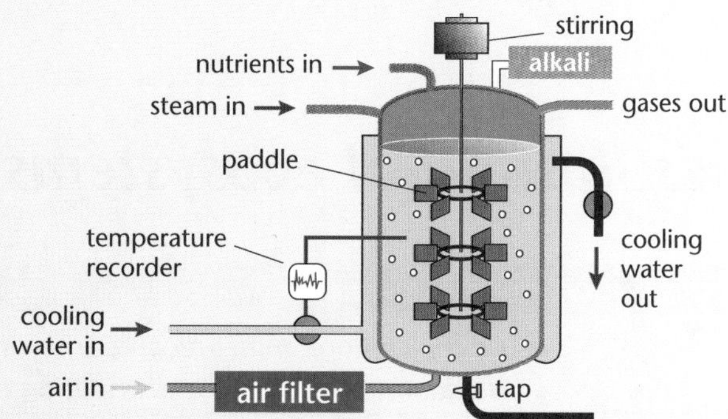

(a) Describe **three** ways in which aseptic conditions are achieved in the fermenter. [3]

(b) If the air filter failed, explain what would be the likely effect inside the fermenter. [3]

[Total: 6]

2 A new genetically modified soya bean plant has been developed. It has a new gene which prevents it from being killed by herbicide (weed killer).

(a) Describe the stages which enable a gene to be transferred from one organism to another. [5]

(b) Suggest how the genetically modified soya plants could result in higher bean yields. [3]

[Total: 8]

3 The graph below shows the level of product secreted by microorganisms in a commercial fermenter.

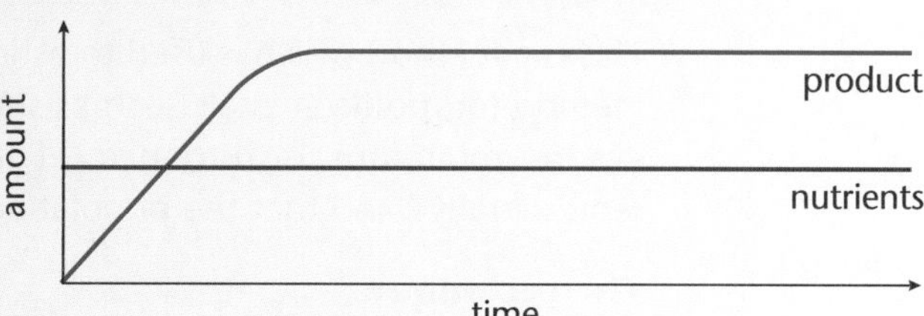

(a) Account for the shape of the graph. [1]

(b) Which type of culture, batch or continuous, took place in this fermenter? Give **two** reasons for your choice. [2]

[Total: 3]

Chapter 7

Ecology and populations

The following topics are covered in this chapter:

- Investigation of ecosystems
- Behaviour
- Colonisation and succession

7.1 Investigation of ecosystems

After studying this section you should be able to:

- use the capture, mark, recapture technique to assess animal populations
- use quadrats to map the distribution of organisms
- understand the factors that affect the distribution of organisms
- describe conservation techniques and methods of population control

LEARNING SUMMARY

Measurement in an ecosystem

OCR 5.3.1-2

The study of ecology investigates the inter-relationships between organisms in an area and their environment. The area in which organisms live is called a **habitat**. The combination of the organisms that live in a habitat and the physical aspects of the habitat is called an **ecosystem**.

Estimating populations

All the individuals of one species living together in a habitat are called a **population**. The size of plant populations can be estimated by using a quadrat placed at random. Animals, however, do not tend to stay still for long enough to be sampled using a quadrat. The population size of an animal species can be estimated by using capture–recapture.

Capture, mark, release, recapture

This is a method which is used to estimate animal populations. It is an appropriate method for motile animals such as shrews or woodlice. The ecologist must always ensure minimum disturbance of the organism if results are to be truly representative and that the population will behave as normal.

Before using the technique you must be assured that:

- there is no significant migration
- there are no significant births or deaths
- marking does not have an adverse effect, e.g. the marking paint should not allow predators to see prey more easily (or vice versa)
- organisms integrate back into the population after capture.

Remember that the method is suitable for large population size only.

The technique

- Organisms are captured, *unharmed*, using a quantitative technique.
- They are counted then discretely marked in some way, e.g. a shrew can be tagged, a woodlouse can be painted (*with non-toxic paint*).
- They are released.
- Organisms from the same population are recaptured, and another count is made, to determine the number of marked animals and the number unmarked.

The calculation

S = total number of individuals in the total population.
S_1 = number captured in sample one, marked and released, e.g. 8.
S_2 = total number captured in sample two, e.g. 10.
S_3 = total marked individuals captured in sample two, e.g. 2.

$$\frac{S}{S_1} = \frac{S_2}{S_3} \quad \text{so, } S = \frac{S_1 \times S_2}{S_3} \quad \text{population} = \frac{8 \times 10}{2} = 40 \text{ individuals}$$

Remember the equation carefully. You will **not** be supplied with it in the examination, but you will be given data.

Measuring the distribution of organisms

This can be measured using another quadrat technique called a belt transect. This method should be used when there is a **transition** across an area, e.g. across a pond or from high to low tide on the sea shore. Use belt transects where there is **change**. The belt transect is a line of quadrats. In each quadrat a measurement such as density can be made. One transect is not enough! Always do a number of transects then find an average for quadrats in a similar zone.

A bar graph would be used to show the **distribution** of plant species across the pond. Note that there would be more than just two species. The graphs show how you could illustrate the data. Clearly flag irises occupy a different niche to water lilies.

A simplified results table

Quadrat no.	*flag iris*	*water lily*
1	10	0
2	7	0
3	1	0
4	0	5
5	0	4
6	0	0
7	0	5
8	0	3
9	0	0
10	1	0
11	8	0
12	4	0

This is just one belt transect. A number would be used and an average taken for each corresponding quadrat.

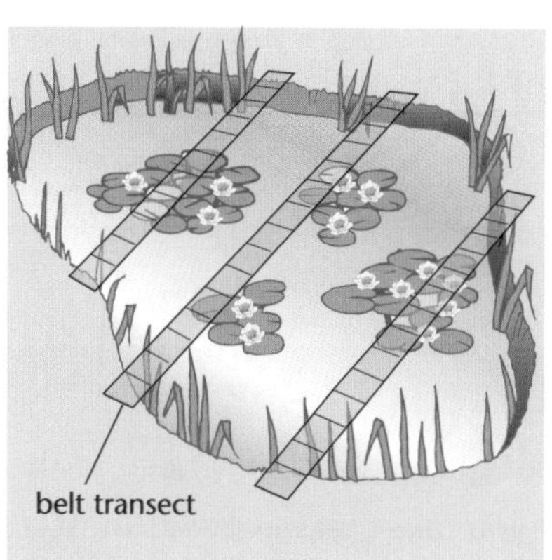

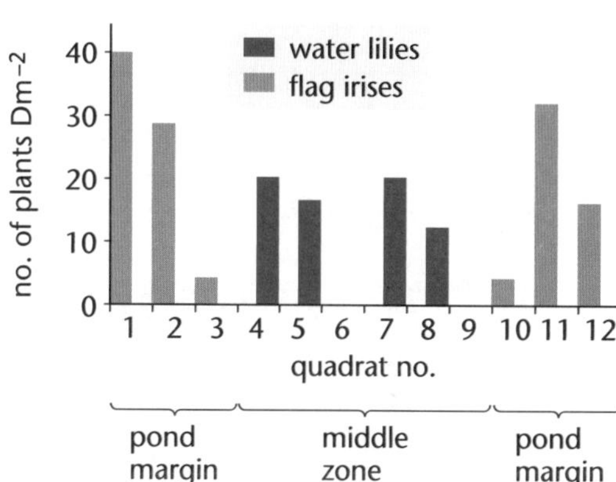

KEY POINT

Other uses of quadrats

Quadrats can also be used to survey animal populations. It is made easier if the organisms are **sessile** (*they do not move from place to place*), e.g. barnacles on a rock. In a pond the belt transect could be coupled with a kick sampling technique. Here rocks may be disturbed and escaping animals noted. Adding a further technique can help, such as using a catch net in the quadrat positions. The principle here is that the techniques are **quantitative**.

Factors that determine population size

Graphical data can show relative numbers and distribution of organisms in a habitat. The ecologist is interested in the factors that determine the size and distribution of organisms.

These factors can be **biotic** or **abiotic**.

Abiotic factors are non-living factors. They include:

- carbon dioxide level
- oxygen level
- pH
- light intensity
- mineral ion concentration
- level of organic material.

Biotic factors are living factors.

They include:

- **Competition** This occurs when organisms are trying to get the same resources. There are two types. **Interspecific competition** takes place when **different** species share the same resources. **Intraspecific competition** takes place when the **same** species share the same resources.
- **Predation** This involves feeding relationships.

Predators and prey

Note that graphs are often given in predator–prey questions. A flush of spring growth is often responsible for the increase in prey. Plant biomass may not be shown on the graph! Candidates are expected to suggest this for a mark. Also remember that as prey increase, their numbers will go down when eaten by the predator. Predator numbers rise after this.

There can be many examples of this type of relationship in an ecosystem. **Primary consumers** rely on the **producers**, so a flush of new vegetation may give a corresponding increase in the numbers of primary consumers. Predators which eat the primary consumers may also follow with a population increase. Each population of the ecosystem may have a **sequential effect** on other populations. Ultimately, the ecosystem is in **dynamic equilibrium** and has limits as to how many of each population can survive, i.e. its **carrying capacity**.

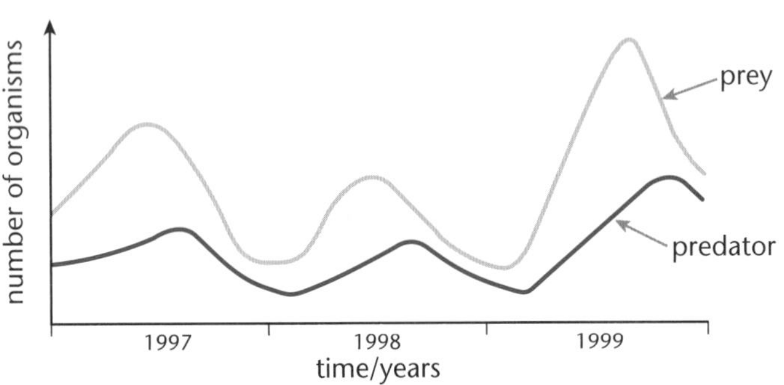

Progress check

(a) List the abiotic factors you may need to measure in a pond survey.

(b) How could you take measurements most efficiently, over a 24-hour period?

(a) Oxygen, carbon dioxide, pH, light, temperature, mineral ions.

(b) Use of environmental probes, interface and computer.

Ecological conservation

OCR 5.3.2

In a world where human population increase is responsible for the destruction of so many habitats, it is necessary to retain as many habitats as possible. Ecological surveys report to governments and difficult decisions are made. Fragile habitats like the bamboo woodlands of China support a variety of wildlife. Conservation areas need to be kept and maintained to prevent extinction of organisms at risk. In the UK we have **sites of special scientific interest** (SSSI) which are given government protection.

Conservation requires management

Although the word 'conservation' implies to 'keep' something as it is, much effort is needed. An area of climax vegetation, e.g. oak woodland, is less of a problem, since it will not change if merely left to its own devices. However, many of the seral stages, e.g. birch woodland along the route to climax, require much maintenance.

Whilst **conservation** usually involves management in order to maintain biodiversity, **preservation** involves protecting areas in their untouched state.

Animal populations need our help, especially when it is often by our own introduction that specific species have colonised an area. Deer introduced into a forest may thrive initially but due to an efficient reproductive rate exceed the carrying capacity of the habitat. **Carrying capacity** is the population of the species which can be adequately supported by the area.

Sometimes herbivores could destroy their habitat by overgrazing, and so must be **culled. Predators** could be introduced to reduce numbers, but they also may need culling at some stage. **Difficult decisions** need to be taken. In the aquatic habitats similar problems exist. Cod populations in the North Sea are being reduced by over-fishing. Agreements have been made by the EU to **reduce fishing quotas** and create **exclusion zones** to **allow fish stocks to recover**. Even before this agreement, smaller fish had to be returned to the sea after being caught to increase the chances of them growing to maturity and breeding successfully.

Endangered species require protection

All over the world many animals and plants are at the limits of their survival. The World Wide Fund for Nature is a charity organisation which helps. The organisation receives support from the public and artists such as David Shepherd. He gives donations from the sale of all of his wildlife paintings, helping to maintain the profile of animals so that we invest in survival projects like protected reserves.

7.2 Colonisation and succession

After studying this section you should be able to:

- *understand how colonisation is followed by changes*
- *understand how colonisation and succession lead to a climax community*

LEARNING SUMMARY

How colonisation and succession take place

OCR 5.3.1

Colonisation and succession also take place in water. Even an artificial garden pond would be colonised by organisms naturally. Aquatic algae would arrive on birds' feet.

Any area which has never been inhabited by organisms may be available for **primary succession**. Such areas could be a garden pond filled with tap water, lava having erupted from a volcano, or even a concrete tile on a roof. The latter may become colonised by lichens. Occasionally an ecosystem may be destroyed, e.g. fire destroying a woodland. This allows **secondary succession** to begin, and signals the reintroduction of plant and animal species to the area.

- **Pioneer species (primary colonisers)** begin to exploit a 'new' habitat. Mosses may successfully grow on newly exposed heathland soil. These are the **primary colonisers** which have adaptations to this environment. Fast germination of spores and the ability to grow in waterlogged and acid conditions aid rapid colonisation. These plants may support a specific food web. In time, as organic matter drops from these herbaceous colonisers it is decomposed, nutrients are added to the soil and acidity increases. In time, the changes caused by the primary colonisers make the habitat unsuitable.
- Conditions unsuitable for primary colonisers may be ideal for other organisms. In heathland, mosses are replaced by heathers which can thrive in acid and xerophytic (desiccating) conditions. This is **succession**, where one community of organisms is replaced by another. In this example, the secondary colonisers have replaced the primary colonisers; this is known as **seral stage 1** in the succession process. Again, a different food web is supported by the secondary colonisers.

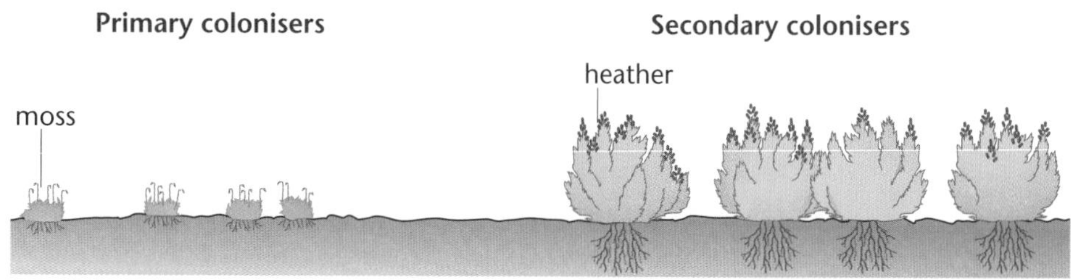

- At every seral stage, there are changes in the environment. The **second seral stage** takes place as the tertiary colonisers replace the previous organisms. In heathland, the new conditions would favour shrubs such as gorse and bilberry plus associated animals.

- The shrubs are replaced in time with birch woodland, the **third seral stage**. Eventually, acidic build up leads to the destruction of the dominant plant species.
- Finally, conditions become suitable for a dominant plant species, the oak. Tree saplings quickly become established. Beneath the oak trees, grasses, ferns, holly and bluebells can grow as a balanced community. This final stage is **stable** and can continue for hundreds of years. This is the **climax community**. Associated animals survive and thrive alongside these plant resources. Insects such as gall wasps exploit the oak and dormice eat the wasp larvae.

Jays are birds which eat some acorns but spread others which they store and forget. The acorns germinate; the woodland spreads.

Climax community

In Britain, an excellent example of a climax community is Sherwood Forest where the 'Major Oak' has stood for 1000 years. Agricultural areas grow crops efficiently by **deflecting succession**. Plants and animals in their natural habitat are 'more than a match' for domesticated crops. Herbicides and pesticides are used to stop the invaders!

7.3 Behaviour

After studying this section you should be able to:

- *describe innate behaviour, kinesis and taxis*
- *understand habituation and imprinting*
- *describe a range of territorial behaviour*

LEARNING SUMMARY

The behaviour of organisms

OCR 5.4.3

Organisms respond to the biotic and abiotic factors of their environment. Biotic factors include response to other species, e.g. the feeding behaviour of grouse from heather on moorlands and the use of the heather to hide from predators. The grouse also respond to each other, e.g. in courtship display. There are different types of behaviour but they are either innate or learned.

Innate behaviour

This behaviour is '**pre-programmed**' by an organism's **genes**. When analysing behaviour it is difficult to determine whether it is innate or learned.

It is safe to say that immediately after the birth of a baby the 'sucking' action to obtain milk from the mother's mammary glands is innate. Similarly, the pecking behaviour of a chicken, while still in an egg, to break the shell, must be innate. As an animal gets older it may well develop patterns of behaviour learned from its experiences. It becomes more and more **difficult to categorise** the behaviour.

Kinesis

This takes place when the response of an organism is **proportional to the intensity of a stimulus**. Kinesis takes the form of an **increase in movement**, but this is **non-directional**. An example of kinesis is shown by woodlice. Intense heat which would harm the woodlice causes them to increase speed and move in random directions. In this way some of the population have a **greater chance of survival** by finding shelter.

Woodlice also respond to a dry environment by increasing random movements but slow down if they reach high humidity.

Taxis

This is a **directional response to a stimulus**. It can be a **positive taxis**, towards, or **negative taxis**, away. An example can be seen using a microscope to observe a group of living specimens of *Euglena viridis*. This is a protoctistan which photosynthesises. Individuals swim to an air bubble and cluster around to obtain maximum CO_2 for photosynthesis. This is **positive chemotaxis** because the organism moves towards the CO_2 source.

Learning

This takes place when an organism changes behaviour as a result of experience within the environment. As a result of the experience, future behaviour becomes modified. For example, a pupil misbehaves and is placed in detention. The pupil learns (hopefully!) that the behaviour should not be repeated. The detention is negative reinforcement. Perhaps positive reinforcement is better to support good behaviour.

Conditioned reflexes

Pavlov experimented with dogs.

- He checked that the group of dogs did not produce saliva when he rang a bell at a time not related to feeding. (**Control**)
- He fed groups of dogs at a specific time each day.
- He measured the amount of saliva produced just before they were fed.
- He then began to ring a bell just before giving the food.
- The dogs began to salivate profusely.
- The bell would elicit exactly the same response as the original stimulus (the food).
- After a while the level of salivation decreased if the food reward was not given.
- Without **positive reinforcement** the level of response would finally disappear completely.

Pavlov's dogs learnt to associate one stimulus to another. This is now described as **classical conditioning**.

If an animal learns to associate an action with getting a reward or punishment then this is called **operant conditioning**.

We are conditioned to respond to advertising in a similar way. A cola drink advertisement uses the latest popular song and glamorous models. We go to the supermarket and respond by buying the product, relating it to the pleasurable experience of the advertisement. Repeat purchases will only continue if the taste of the cola elicits a positive taste perception.

Habituation

This takes place when an organism is subjected to a **stimulus which is not harmful or rewarding**. As a result of continued subjection to a stimulus a **response will gradually decrease** and can finally disappear completely. A farmer puts an electronic bird scarer into a field. Birds are frightened off by frequent 'bangs'. They return, gradually getting closer and finally learn that the scarer is non-threatening. Soon they feed close to the scarer which has no effect. This is **habituation**.

Advertisements have a short 'shelf-life'. Continued exposure to the same advertisement results in habituation so that the response decreases. No wonder media advertising is replaced every few weeks!

The Sand Hill crane and imprinting
This endangered species is reared in incubators and re-introduced into the wild. There is a problem, though. Young cranes would imprint upon humans, so when re-introduced into the wild, it would move towards people. This would be dangerous, so each day, keepers dress up in 'crane' uniforms. In the wild the birds then move towards groups of adult cranes.

Imprinting

This takes place during the very early life of an organism, e.g. a chick emerges from its egg shell and immediately **bonds** with an object close by. In nature, this will normally be the mother. The mother will impart useful behavioural patterns to the chick, thus having **survival value**. From an incubator, the focus of the imprinting would be a human. The imprinting behaviour is that the chick, in this instance, will follow the human or any object to which it is first exposed.

Latent learning

If an animal encounters a new habitat it will investigate the area, learning to find its way around. This information is not of immediate use but may become useful in the future if for example, it is surprised by a predator. This type of learned behaviour is called latent learning.

Insight learning

Insight learning is the most advanced form of learning and it involves animals being able to predict the outcome of their actions. One of the most famous sets of observations was made on chimpanzees. The chimps could not reach some bananas that were outside their cage, even when provided with sticks. The chimps then learned to join the sticks together to enable them to reach the bananas.

Territorial behaviour

Populations of organisms living in an area can benefit from territorial behaviour. Too many animals of the **same species**, living in an area, **competing** for the same food would put the whole population in danger. Many species display territorial behaviour which prevents this outcome.

Examples are given to illustrate **principles.** It is unlikely that you will be given the same examples in your examination. Apply the principles to the given data.

Defence of the territory

- Animals are often **aggressive to members of the same species**, outside of the same family group.
- Territory is demarcated in a variety of ways, such as marking with **urine**, **faeces** or **scent**. Birds use **song**, whereas other animals have characteristic **calls.** Excluding others in these ways **can prevent physical confrontation** which often results in injury.
- It is an advantage to the species to have a **feeding range** which excludes others. This increases the chances of there being **enough food for the family group**.
- The apportioning of territories serves as **density dependent regulation** so that the best use is made of existing resources.
- A further advantage is that a territory marks out a designated **mating area**. Other males will usually remain outside the zone. Offspring have protection for their early development.

What is the advantage of male aggression to other males in a population?
The fittest organisms need to pass on their advantageous genes to offspring. In deer herds, a dominant stag (male) is challenged by a younger male occasionally. Antler to antler fights take place and there is potential damage to both stags. In time a new dominant stag takes over the family group and now has exclusive mating rights with a group of hinds (females). This behaviour ensures the male reproductive role involves only the strongest males. The gene pool is improved!

Courtship behaviour

This behaviour is species dependent and courtship display is anchored in the genes. Courtship rituals are very important to ensure that:

- the opposite sexes recognise each other
- the animals will mate with organisms from the same species (mating will produce fertile offspring)
- the act of mating is synchronised with the oestrous cycle. In pigs, a boar is always ready to mate but a sow is only receptive to him at ovulation. She produces pheromone attractants to encourage the boar.

Sample question and model answer

When given a passage, line numbers are often referred to. Try to understand the words in context. Do not rush in with a pre-conceived idea.

Read the passage, then answer the questions below.

Around the UK coast there are two species of barnacle, *Chthamalus stellatus* and *Balanus balanoides*. Both species are sessile, living on rocky sea shores.

The adult barnacles do not move from place to place but do reproduce sexually. They use external fertilisation. Larvae resemble tiny crabs and are able to swim. At a later stage these larvae come to rest on a rock where they become fixed for the remainder of their lives.

The barnacles are only able to feed while submerged.

Adult *Chthamalus* are found higher on the rocks than *Balanus* in the adult form as shown in the diagram below. Scientists have shown that the larvae of each species are found at all levels.

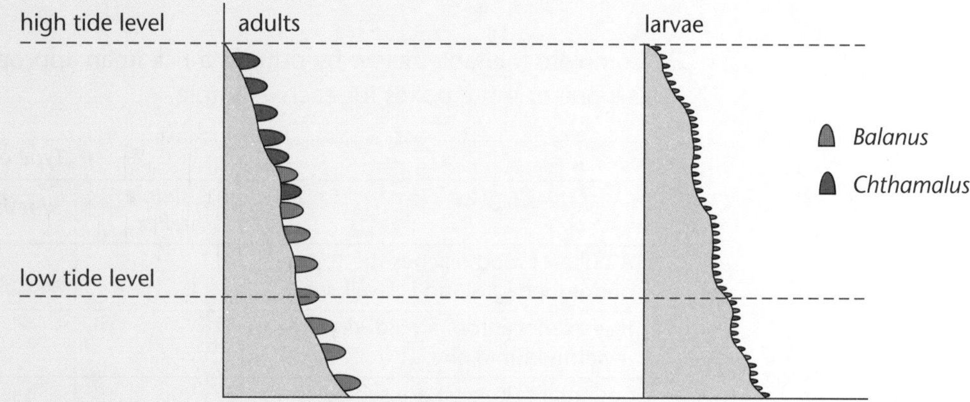

In this question are key terms, of which you will need to recall the meaning. This is only possible with effective revision. Did you already know the key terms **genus**, **sessile** and **motile**? Knowledge of these terms would enable you to access other marks, only easy when you have key word understanding. Try writing out a glossary of terms to help your long-term memory.

(a) Name the genus for each barnacle. [1]

Chthamalus and Balanus

(b) What does the term sessile mean? (line 2) [1]

is not motile, i.e. does not move from place to place

(c) Suggest how it is possible for neighbouring *Balanus* individuals to breed sexually (line 5) with each other even though they are sessile. [1]

produce sperms which swim through the water

(d) Explain **one** advantage of the larvae being motile. [2]

able to colonise new areas

where there may be more nutrients / food

(e) Which type of competition exists between *Chthamalus* and *Balanus*? [1]

interspecific competition

(f) Suggest an explanation for the distribution of each species of barnacle. [6]

The larvae are found at all tide levels; at a lower tide level the barnacles are submerged for longer; Balanus may grow at a faster rate; can compete for food better; Chthamalus, which die out at lower levels near higher tide level the barnacles are exposed to open air for longer; Balanus may not be adapted to withstand desiccation; whereas Chthamalus can withstand drying out; and so survive without the competition from Balanus.

Practice examination questions

1 Ecologists wished to estimate the population of a species of small mammal in a nature reserve.

- They placed humane traps throughout the reserve and made their first trapping on day one, capturing 16 shrews.
- They were tagged then released.
- After day four a second trapping was carried out, capturing 12 shrews.
- Five of these shrews were seen to be tagged.

(a) The ecologists must be satisfied of a number of factors before using the 'capture, mark, release, recapture' method. List three of these factors. [3]

(b) Use the data to estimate the shrew population.
Show your working. [2]

(c) Comment on the *level* of reliability of your answer. [1]

[Total: 6]

2 Complete the table below by putting a tick in an appropriate box. You may tick one or more boxes for each example.

	Type of behaviour			
	simple reflex	***kinesis***	***positive taxis***	***negative taxis***
A bolus of food reaches the top of our oesophagus and is swallowed.				
Insects move from a cold, dry area to a warm, humid one.				
Springtails (insects) are subjected to increasingly hot conditions, and react by increasing speed in a number of directions. Some go towards the heat source and die.				
A motile alga swims towards light.				

[5]

[Total: 5]

3 A grebe is a water bird which displays a distinctive courtship ritual.
Male behaviour is distinctive from that of the female.

State **three** advantages to the species of this behaviour. [3]

[Total: 3]

Practice examination questions (continued)

4 The diagrams show stages in the development of a garden pond over a 10-year period.

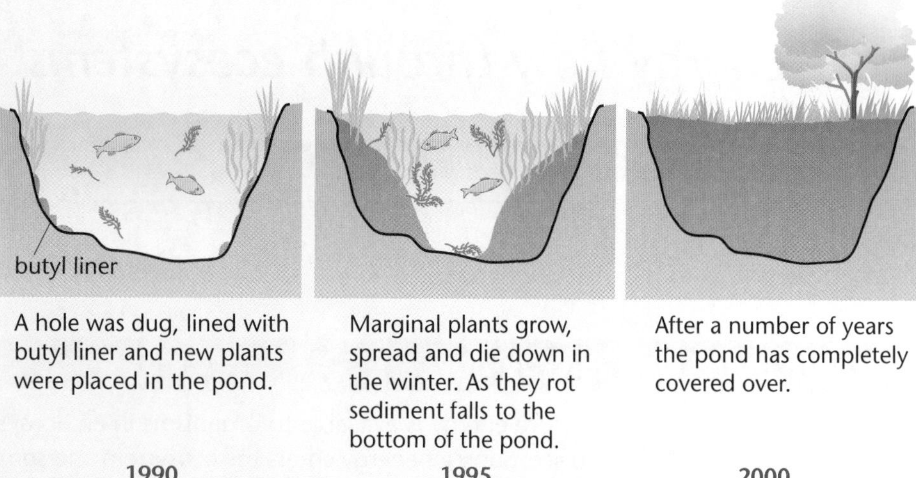

(a) In 1990 irises, oxygenating pondweed and a water lily were planted in the pond. Algae were not planted but arrived in the pond in some other way.

(i) What term describes an organism that grows in a new habitat that previously supported no life? [1]

(ii) After a time the algae produced a thick 'carpet' of growth on the surface of the pond. Explain the effect this may have on organisms under the water. [5]

(b) Describe the stages which took place to produce the stable grassland after 10 years. [4]

[Total: 10]

Chapter 8

Energy and ecosystems

The following topics are covered in this chapter:

- *Energy flow through ecosystems*
- *Nutrient cycles*

8.1 Energy flow through ecosystems

After studying this section you should be able to:

- *understand the roles of producers, consumers and decomposers in food chains*
- *understand the flow of energy through an ecosystem*

LEARNING SUMMARY

Food chains and energy flow

OCR 5.3.1

Before energy is available to organisms in an ecosystem, photosynthesis must take place. Sunlight energy enters the ecosystem and some is available for photosynthesis. Not all light energy reaches photosynthetic tissues. Some totally misses plants and may be absorbed or reflected by items such as water, rock or soil. Some light energy which does reach plants may be reflected by the waxy cuticle or even miss chloroplasts completely! The energy that is trapped by photosynthesis and converted into biomass is called the **gross primary productivity** (GPP).

Around 4% of light entering an ecosystem is actually used in photosynthesis.

KEY POINT

The green plant uses the **carbohydrate** as a first stage substance and goes on to make **proteins** and **lipids**. Plants are a rich source of nutrients, available to the herbivores which eat the plants. Some energy is not available to the herbivores because green plants **respire** (releasing energy).

The energy that is available to herbivores is called the **net primary productivity** (NPP). It is calculated as follows:

Net primary productivity = gross primary productivity – energy lost in respiration

Energy is also lost from the food chain as **not all parts** of plants may be **consumed**, e.g. roots.

Food chains and webs

Energy is passed along a food chain. Each food chain always begins with an **autotrophic** organism (producer), then energy is passed to a primary consumer, then a secondary consumer, then a tertiary consumer and so on.

direction of energy flow →

Producer → primary consumer → secondary consumer → tertiary consumer
(herbivore) (1st carnivore) (2nd carnivore)

Note that a small bird is a secondary consumer when it eats apple codling moths but a tertiary consumer when it eats greenfly.

The following example shows three food chains linked to form a food web.

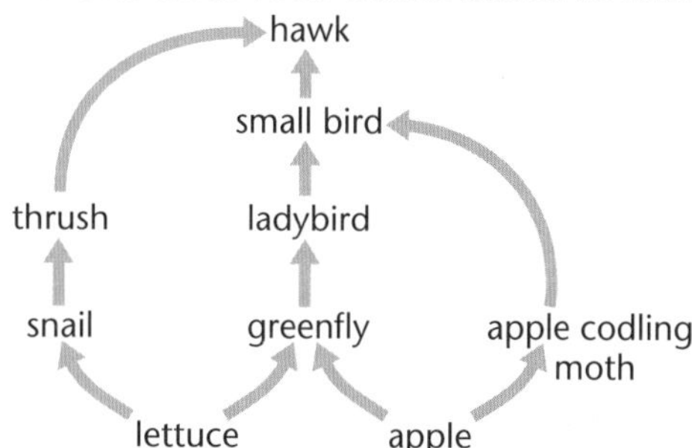

The producers always have more energy than the primary consumers, the primary consumers more than the secondary consumers and so on, up the food web. Energy is released by each organism as it respires. Some energy fails to reach the next organism because not all parts may be eaten.

Each feeding level along a food chain can also be represented by a **trophic level**. The food chain below is taken from the food web above and illustrates trophic levels. Energy may be used by an organism in a number of different ways:

- respiration releases energy for movement or maintenance of body temperature, etc.
- production of new cells in growth and repair
- production of eggs
- released trapped in excretory products.

Examiner's tip! Note that the primary consumer is at trophic level 2. It is easy to make a mistake with this concept. Many candidates do!

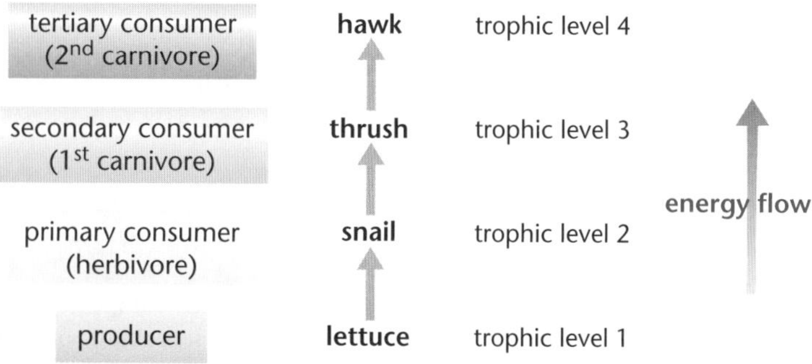

Pyramids of numbers, energy and biomass

A food chain gives limited information about feeding relationships in an area. Actual proportions of organisms in an area give more useful data. Consider a food chain from a wheat field. The pyramid of numbers sometimes does not give a suitable shape. In the example shown below, there are more aphids in the field than wheat plants. This gives the shape shown below (not a pyramid in shape!). A pyramid of biomass is more likely to be a pyramid in shape because it takes into account the size of the organism. It does not always take into account the rate of growth and so only a pyramid of energy is always the correct shape.

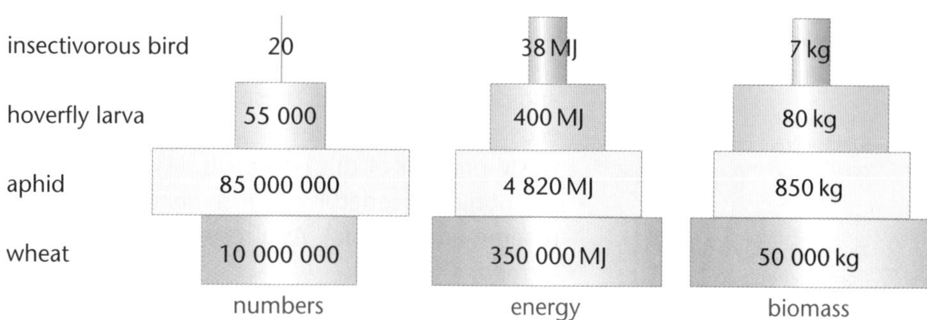

Biomass is the mass of organisms present at each stage of the food chain. The biomass of wheat would include leaves, roots and seeds. (All parts of the plant are included in this measurement.)

The organisms in the above food chain may die rather than be consumed. When this happens, the decomposers use extra-cellular enzymes to break down any organic debris in the environment. Corpses, faeces and parts that are not consumed are all available for decay.

8.2 Nutrient cycles

After studying this section you should be able to:

- *recall how nitrogen is recycled*

LEARNING SUMMARY

The nitrogen cycle

OCR 5.3.1

Nitrogen is found in every amino acid, protein, DNA and RNA. It is an essential element! Most organisms are unable to use atmospheric nitrogen directly so the nitrogen cycle is very important.

There are three parts of the nitrogen cycle which are regularly examined:

- nitrogen fixation in leguminous plants
- nitrification
- denitrification.

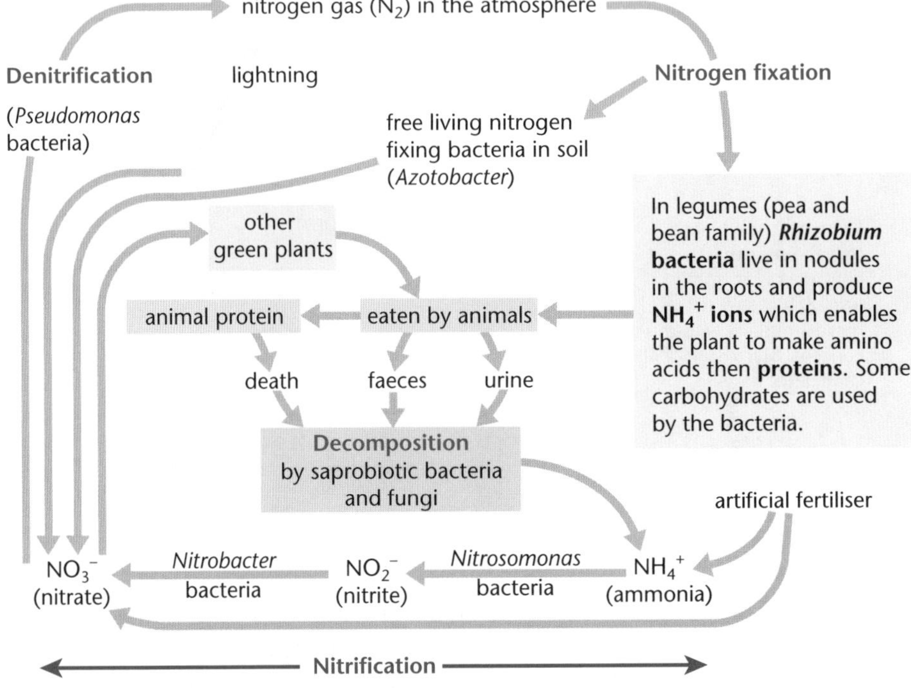

Some important points

The association of *Rhizobium* bacteria with legume plants give advantages to both organisms. This relationship is known as **mutualism**.

Saprobiotic bacteria and fungi secrete extra-cellular enzymes.

The biochemical route from ammonia to nitrate is **nitrification**. This is helped by ploughing which allows air into the soil. Nitrifying bacteria are aerobic. Draining also helps.

- **Nitrogen gas** from the atmosphere is used by ***Rhizobium* bacteria**. These bacteria, living in nodules of legume plants, convert nitrogen gas into **ammonia** (NH_3) then into amine ($-NH_2$) compounds. The plants transport the amines from the nodules and make amino acids then proteins. *Rhizobium* bacteria gain carbohydrates from the plant, therefore each organism benefits.
- Plants support food webs, throughout which excretion, production of faeces and death take place. These resources are of considerable benefit to the ecosystem, but first **decomposition** by **saprobiotic** bacteria takes place, a waste product of this process is **ammonia**.
- Ammonia is needed by *Nitrosomonas* bacteria for a special type of nutrition (chemo-autotrophic). As a result another waste product, **nitrite** (NO_2) is formed.
- Nitrite is needed by *Nitrobacter* bacteria, again for chemo-autotrophic nutrition. The waste product from this process is **nitrate**, vital for plant growth. Plants absorb large quantities of nitrates via their roots.
- Nitrogen gas is returned to the atmosphere by **denitrifying bacteria** such as *Pseudomonas*. Some nitrate is converted back to nitrogen gas by these bacteria. The cycle is complete!

Sample question and model answer

Some questions give information that you can use to deduce the answers (a) (i) is one of them!

This question targets the nitrogen cycle. Be ready to answer questions about *any part* of the cycle. Pure recall of the cycle is not enough! You need to apply your knowledge.

(a) The sequence below shows how nitrate can be produced from a supply of oak leaves.

	decomposers		*Nitrosomonas* bacteria		*Nitrobacter* bacteria	
dead oak leaves	⟶	NH_3 (ammonia)	⟶	NO_2 (nitrite)	⟶	NO_3 (nitrate)

(i) Suggest the consequences of death of the *Nitrosomonas* bacteria. [4]

build up of ammonia; build up of dead leaves; death of ***Nitrobacter*** bacteria; no nitrite/no nitrate

(ii) Name the process by which bacteria produce nitrate from ammonia. [1]

nitrification

(iii) Name **two** populations of organisms not shown in the sequence which would be harmed by a lack of nitrate. [2]

denitrifying bacteria or ***Pseudomonas***; plants or producers

(iv) Which organisms fix atmospheric nitrogen on the nodules of bean plants? [1]

(*Rhizobium*) bacteria

(b) (i) A farmer rears pigs by a factory-farming method. Pigs are kept indoors 24 hours per day in warm, confined cubicles.
How can this method result in the production of a greater yield of pork than from animals reared outside? [3]

less energy is released for movement; less energy is used to maintain body temperature; more energy is used for biomass

(ii) Suggest why many consumers object to this factory-farming method. [1]

cruel or not ethical

Practice examination questions

1 The number of species of grass and the number of leguminous plants growing in two fields was measured over a 10-year period. Field A was given nitrogenous fertiliser each year, but field B was given none. The results are shown in the graphs.

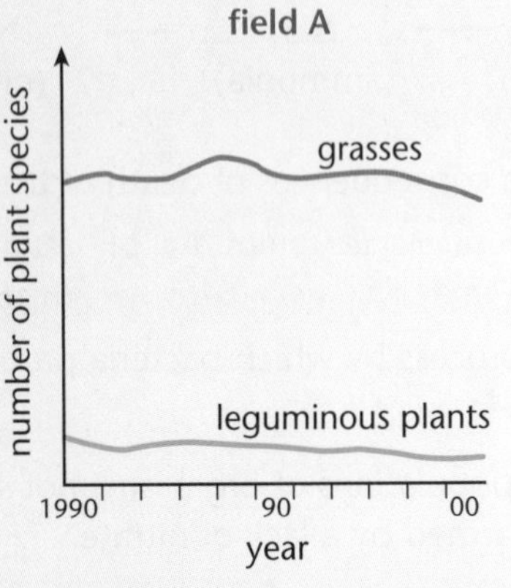

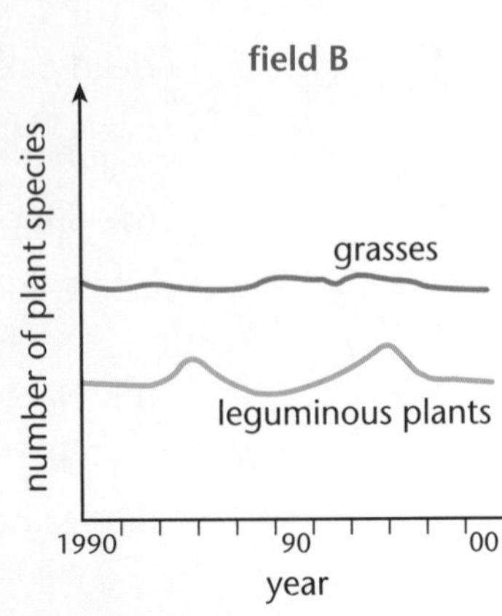

(a) (i) Suggest why there were fewer leguminous plant species in field A. [2]

(ii) Suggest why there were more leguminous plant species in field B. [2]

(b) After the main investigation no fertiliser at all was used in either field. Cattle were allowed to graze in both fields. At the end of five years the number of legume species in each field had decreased. Suggest why the number of legume plants decreased. [1]

[Total: 5]

2 The diagram shows the flow of energy through an ecosystem. The energy units are in kJ.

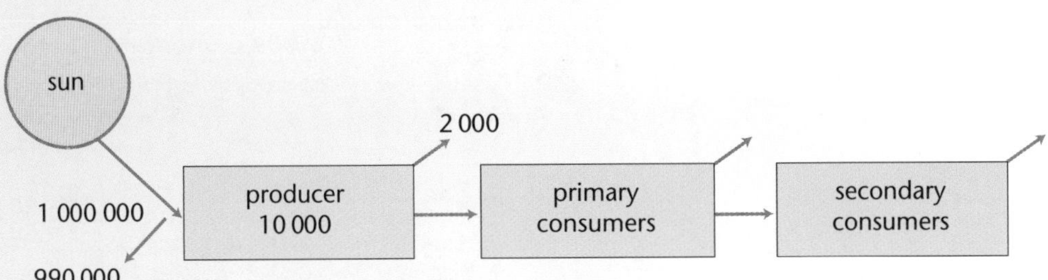

(a) Much of the light energy falling on the producers is not used for photosynthesis. Give two reasons why. [2]

(b) The GPP of the producers is 10 000kJ

(i) What is their NPP? [1]

(ii) Why is there a difference between their GPP and their NPP? [1]

(c) Explain why most food chains are limited to about five trophic levels. [3]

[Total: 7]

Chapter 9

Synoptic assessment

What is synoptic assessment?

You must know the answer to this question if you are to be fully prepared for your A2 examinations!

Synoptic assessment:

- involves the drawing together of knowledge, understanding and skills learned in different parts of the AS/A2 Biology courses
- requires that candidates apply their knowledge of a number of areas of the course to a variety of contexts
- is tested at the end of the A2 course both by assessment of investigative/ practical skills and by examinations
- is valued at 20% of the marks of the course total.

In the OCR examinations there will be synoptic question on both Unit 4 and Unit 5 papers.

Practical investigations

You will need to apply knowledge and understanding of the concepts and principles, learned throughout the course, in the **planning**, **execution**, **analysis** and **evaluation** of each investigation.

How can I prepare for the synoptic questions?

- **Check out the modules** which will be examined for the synoptic questions.
- Expect **new contexts** which draw together lots of different ideas.
- Get ready to **apply** your knowledge to a new situation; contexts change but the **principles remain the same**.
- In modular courses there is sometimes a tendency for candidates to learn for a module, achieve success, then forget the concepts. Do not allow this to happen! **Transfer concepts** from one lesson to another and from one module to another. Make those connections!
- Improve your powers of analysis – **take a range of different factors into consideration** when making conclusions; synoptic questions often involve both graphical data and comprehension passages.
- Less able candidates make limited conclusions; high ability candidates are able to consider several factors at the same time, then make a **number of sound conclusions** (not guesses!).
- You need to do **regular revision** throughout the course; this keeps the concepts 'hot' in your memory, 'simmering and distilling', ready to be **retrieved** and **applied** in the synoptic contexts.
- The bullet point style of this book will help a lot; back this up by summarising points yourself as you make notes.

Why are synoptic skills examined?

Once studying at a higher level or in employment, having a narrow view, or superficial knowledge of a problem, limits your ability to contribute. Having discrete knowledge is not sufficient. You need to have confidence in applying your skills and knowledge.

Synoptic favourites

The final modules, specified by OCR for synoptic assessment, include targeted synoptic questions. Concepts and principles from earlier modules will be tested together with those of the final modules. You can easily identify these questions, as they will be longer and span wide-ranging ideas.

Can we predict what may be regularly examined in synoptic questions?

'Yes we can!' Below are the top five concepts. Look out for common processes which permeate through the other modules. An earlier module will include centrally important concepts which are important to your understanding of the rest.

KEY POINT

Check out the synoptic charts

1 Energy release

Both aerobic and anaerobic respiration release energy for many cell processes. Any process which harnesses this energy makes a link.

Synoptic links

Try this yourself! Think logically. Write down an important biological term such as 'cell division'. Link related words to it in a 'flow diagram' or 'mind map'. The links will become evident and could form the framework of a synoptic question.

Examples

- Reabsorption of glucose involves active transport in the proximal tubule of a kidney nephron. If you are given a diagram of tubule cells which show both mitochondria and cell surface membrane with transporter proteins, then this is a cue that active transport will probably be required in your answer.
- Contraction of striated (skeletal) muscle requires energy input. This is another link with energy release by mitochondria and could be integrated into a synoptic question.
- The role of the molecule ATP as an energy carrier and its use in the liberation of energy in a range of cellular activity may be regularly linked into synoptic questions. The liberation of energy by ATP hydrolysis to fund the sodium pump action in the axon of a neurone.
- The maintenance of proton gradients by proton pumps is driven by electron energy. Any process involving a proton pump can be integrated into a synoptic question.

Energy: input and output

This has to be a favourite for many synoptic questions. Energy is involved in so many processes that the frequency of examination will be high.

2 Energy capture

Photosynthesis is responsible for availability of most organic substances entering ecosystems. It is not surprising that examiners may explore knowledge of this process and your ability to apply it to ecological scenarios.

Examples

- Given the data of the interacting species in an ecosystem you may be given a short question about the mechanism of photosynthesis then have to follow the energy transfer routes through food webs.
- Often both photosynthesis and respiration are examined in a synoptic type question. There are similarities in both the thylakoid membranes in chloroplasts and cristae of mitochondria.
- Many graphs in ecologically based questions show the increase in herbivore numbers, followed by a corresponding carnivore increase. Missed off the graph, your knowledge of a photosynthetic flush which stimulates herbivore numbers may be expected.

DNA: fundamental to life

A high profile molecule involved in many biochemical and biotechnological processes. It must figure regularly in synoptic questions. Genetic engineering, the process, and how it can be harnessed to solve problems may be regularly tested.

3 The structure and role of DNA

It is important to know the structure of DNA because it is fundamentally important to the maintenance of life processes and the transfer of characteristics from one generation of a species to the next. DNA links into many environmentally and evolutionally based questions.

- The ultimate source of variation is the mutation of DNA. Questions may involve the mechanism of a mutation in terms of DNA change and be followed by natural selection. This can lead to extinction or the formation of a new species. Clearly there are many potential synoptic variations.
- DNA molecules carry the genetic code by which proteins are produced in cells. This links into the production of important proteins. The structure of a protein into primary, secondary, tertiary and quaternary structure may be tested. All enzymes are proteins, so a range of enzymically based question components can be expected in synoptic questions.
- The human genome project is a high-profile project. The uses of this human gene 'atlas' will lead to many developments in the coming years. The reporting of developments, radiating from the human genome project, could be the basis of many comprehension type questions, spanning diverse areas of Biology. Save newspaper cuttings, search the internet and watch documentaries. Note links with genetic diseases, ethics, drugs, etc.

4 Structure and function of the cell surface membrane

There are a range of different mechanisms by which substances can cross the cell surface membrane. These include diffusion, facilitated diffusion, osmosis, active transport, exocytosis and pinocytosis. Additionally glycoproteins have a cell recognition function and some proteins are enzymic in function. Knowledge of these concepts and processes can be tested in cross-module questions.

- In an ecologically based question the increasing salinity of a rock pool in sunny conditions could be linked to water potential changes in an aquatic plant or animal. Inter-relationships of organisms within a related food web could follow, identifying such a question as synoptic.
- In cystic fibrosis a transmembrane regulator protein is defective. A mutant gene responsible for the condition codes for a protein with a missing amino acid. This can link to the correct functioning of the protein, the mechanism of the mutation and the functioning of the DNA.

Synoptic predictions

The list is given as an attempted prediction. There may be other links not listed in this chapter! Look at the specification. You will be given more detailed information. Check past papers for OCR as the pattern is created. Do not be phased by new ideas and unknown organisms. Simply apply the concepts and principles learned in the course to new situations to be successful.

5 Transport mechanisms

This theme may unify the following into a synoptic question, transport across membranes, transport mechanisms in animal and plant organs. Additionally, they may be linked to homeostatic processes.

- The route of a substance from production in a cell, through a vessel to the consequences of a tissue which receives the substance, could expand into a synoptic question. Homeostasis and negative feedback could well be linked into these ideas.

Sample questions and model answers

A short question which cuts across the course. It refers back to AS. Do not forget those modules! See the AS Guide for additional advice and those concepts not found in A2.

Question 1 *(a short structured question)*

The kangaroo rat (*Dipodomys deserti*) is a small mammal that lives in the Californian desert. It has specialised kidneys so that it can produce a very concentrated urine.

(a) Name the genus that contains the kangaroo rat. [1]

Dipodomys

(b) What is the biological naming system called that gives the kangaroo rat its scientific name? [1]

binomial system

(c) Kangaroo rats have long loops of Henle. In which part of the kidney would you expect to find loops of Henle? [1]

medulla

(d) What is the name of the hormone that controls the concentration of the urine in mammals? [1]

ADH

(e) Which gland releases this hormone into the blood? [1]

pituitary gland

(f) The desert community that contains the kangaroo rat is the final product of succession in California. What is the name of the final, stable community that is produced by succession? [1]

a climax community

[Total: 6]

Question 2 *(a longer, more open-ended question)*

Plants and animals both need to exchange gases with the environment. Describe how animals and plants are adapted for efficient gaseous exchange. [10]

(Quality of written communication assessed in this answer.)

- examples of respiratory surfaces in animals:
  gills/lungs;
  tracheoles in insects;
  surface of protoctists;
  stomata in plants
- large surface area:
  way(s) in which this is achieved e.g.
  many alveoli;
  surface area/volume ratio in protoctists;
  many gill filaments;
  large surface area of leaves;
  many mesophyll cells
- maintenance of diffusion gradients
  way(s) in which this is achieved
  rich blood supply;
  ventilation mechanisms;
  sub-stomatal airspaces;
  spongy mesophyll air spaces;
  use of carbon dioxide in mesophyll cells

Sample questions and model answers (continued)

- small diffusion pathway
  barriers one cell thick;
  specialised cells, e.g. squamous epithelium;
  thin cell walls of palisade cells

Note, there is one mark available for legible text with accurate spelling, punctuation and grammar.

[Total: 10]

Question 3 *(a longer question of higher mark tariff)*

Prepare yourself for this type of synoptic question. It cuts across a large part of the specification. Make the links with different ideas. This fact is very important; concepts from AS are needed.

Different concentrations of maltose were injected in the small intestine of a mouse. The amount of glucose appearing in the blood and the small intestine after 15 minutes were measured. The results are shown in the graph.

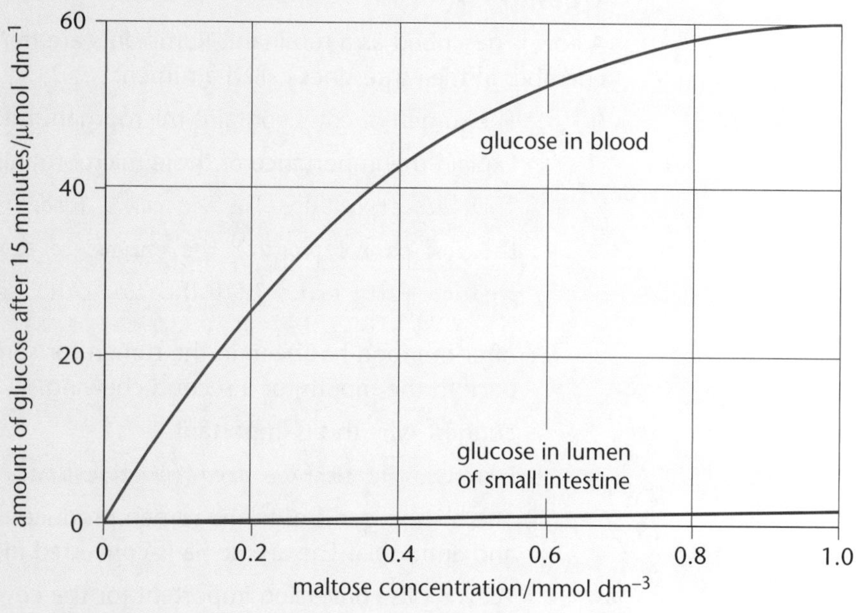

(a) (i) Describe the structure of a maltose molecule [2]

Two molecules of (alpha) glucose;
joined together by a glycosidic bond.

(ii) Maltose is converted into glucose by a hydrolysis reaction. What is a hydrolysis reaction? [1]

A reaction that breaks down a substance by the addition of water.

(b) Describe the effect of different maltose concentrations on the amount of glucose found in the lumen of the small intestine compared to the effect on the amount found in the blood. [2]

the maltose concentration has much more effect on the amount of glucose in the blood; the amount of glucose found in the blood is starting to level off but the amount in the lumen is increasing steadily.

Even if you only cover one of these points, you can pick up a second mark by correctly using figures from the graph in your answer.

Sample questions and model answers (continued)

This is a harder stretch and challenge question.

(c) The enzyme maltase is found on the cell surface membrane of the epithelial cells of the small intestine.

(i) How does the data on the graph indicate that the enzyme is not released into the lumen? [1]

Very little/no increase in the amount of glucose in the lumen.

(ii) Explain why having the enzyme fixed to the cell surface will increase the rate of glucose absorption. [2]

Higher concentration of glucose produced close to intestinal lining;

will increase the concentration gradient between intestine and blood.

[Total: 8]

Question 4

A cow is described as a ruminant. Ruminants are herbivores that have a chamber in their intestines called a rumen.

(a) (i) The rumen of cows contains microorganisms.

Explain the importance of these microorganisms to the cow. [3]

They digest cellulose in the cow's food;

the cow cannot produce the enzyme to digest cellulose;

produce fatty acids that the cow can use.

(ii) After the food has been in the rumen for some time it is regurgitated back to the mouth for a second chewing.

Suggest why this is important. [1]

Increase the surface area for digestion.

(iii) The microorganisms in the rumen produce two waste products, methane and ammonia. The ammonia is converted into urea by the cow's liver.

Why is this conversion important for the cow? [1]

Ammonia is more toxic than urea.

(b) The table shows the amount of methane produced by different domesticated animals

Animal type	*Methane production per animal in kg per animal per year*	*Total methane production in tonnes per year*
buffaloes	50	6.2
camels	58	1.0
goats	5	2.4
sheep	6	3.4

(i) Which of the animals in the table are ruminants?

Explain how you can tell this. [2]

Buffaloes and camels;

they produce much more methane per animal.

Sample questions and model answers (continued)

This is a typical synoptic question as it links two different topics, digestion in herbivores and the greenhouse effect!

(ii) Which type of animal in the table is domesticated in the highest numbers? Explain how you worked out your answer. [2]

Sheep; dividing the total methane production by the production per animal gives the highest number.

(iii) Methane is a potent greenhouse gas. What is a greenhouse gas? [2]

A gas that prevents the escape of infra red radiation from the atmosphere; therefore causes the atmosphere to warm.

(iv) It has recently been discovered that methane is released when arctic ice melts. Explain why people are concerned by this discovery. [2]

The release of methane would increase global warming;
which in turn would result in the release of even more methane.

[Total: 13]

Practice examination answers

Chapter 1 Energy for life

1

(a) in cytoplasm [1]

(b) pyruvate [1]

(c) 2 ATPs begin the process;
2ATPs are produced from each of the two GP molecules, so –2 + 4 = +2 ATPs net [1]

(d) animal; animal cells produce lactate [1]

(e) oxygen or aerobic [1]

[Total: 5]

2

(a) At this point the amount of carbon dioxide given off by the plant in *respiration*, is totally used by the plant in *photosynthesis*. [2]

(b) compensation point [1]

(c) The continued graph line falls (as light dims); line ends below the horizontal axis (when it's dark!). [2]

[Total: 5]

3

(a) Absorption spectrum is obtained from the amount of each wavelength absorbed by the pigments which made up the chlorophyll of the plant.

Action spectrum is produced by measuring the amount of photosynthesis by the plant for each separate wavelength. [2]

(b) Low amount of photosynthesis because not much light energy absorbed, most is reflected. [1]

(c) Evolution of oxygen, collected by water displacement. [1]

[Total: 4]

4

(a) mitochondrion [1]

(b) NADH [1]

(c) cytochrome [1]

(d) ATP [1]

[Total: 4]

5

(a) (i) rate of photosynthesis is proportional to light intensity; rate limited by amount of light available

(ii) as light intensity increases it results in significantly less increase in the rate of photosynthesis

(iii) rate of photosynthesis has levelled off, no longer limited by light (but other conditions could be limiting!). [3]

(b) Similar shape of graph, begins at origin, but graph line above the given plotted curve. [1]

[Total: 4]

Chapter 2 Response to stimuli

1

(a) (i) IAA (at these lower) concentrations is *proportional* to the angle of curvature of the stem. [1]

(ii) IAA (at these higher) concentrations is *inversely proportional* to the angle of curvature. [1]

(b) *More* IAA causes the cells at side of stem in contact with agar block to elongate more than other side.

So this side grows more strongly bending stem towards the weaker side. [2]

(c) Growth is only stimulated up to a certain high IAA concentration, after this curvature would be inhibited. [2]

[Total: 6]

2

(a) A = actin
B = myosin [2]

(b) action potential reaches sarcomere [1]

(c) both filaments slide alongside each other;
they form cross bridges;
during contraction the filaments slide together to form a shorter sarcomere [2]

[Total: 5]

3

(i) resting potential achieved; [2]
Na^+ / K^+ pump is on

(ii) Na+ / K^+ pump is off;
so Na^+ ions enter axon [2]

(iii) maximum depolarisation achieved; K^+ ions leave [2]

(iv) Na^+ ions leave due to Na^+/ K^+ pump being back on;
this is during the refractory period;

(v) at end of this resting potential re-established;
axon membrane re-polarised [4]

[Total: 10]

Chapter 3 Homeostasis

1 (a)

	Nervous system	*Endocrine system*
Usually have longer lasting effects		✓
Have cells which secrete transmitter molecules	✓	
Cells communicate by substances in the blood plasma		✓
Use chemicals which bind to receptor sites in cell surface proteins	✓	✓
Involve the use of Na^+ and K^+ pumps	✓	

[2]

(b) homeostasis [1]

[Total: 3]

2

It increases permeability of; the collecting ducts, and the distal convoluted tubules of the nephron;

- more water drawn out of the collecting ducts;
- by the sodium and chloride ions;
- in medulla of kidney;
- so more water can be reabsorbed back into blood;
- through the capillary network. (max 6) [6]

[Total: 6]

3

(a)

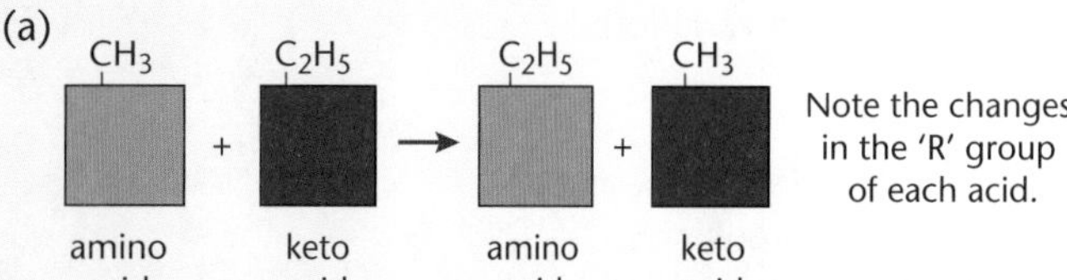

[2]

(b) (i) liver [1]

(ii) To make different amino acids with the help of the essential amino acids. [2]

[Total: 5]

4

(a) **B**, because as glucose levels rose after meals they did not decrease enough (this kept the blood glucose level too high) [1]

(b) glucose levels fell after every meal, so glucose must have entered the cells and liver [1]

(c) in the pancreas;
in the β cells of islets of Langerhans (max 2) [2]

[Total: 4]

Chapter 4 Further genetics

1

(a) no immigration and no emigration; no mutations; no natural selection; true random mating; all genotypes must be equally fertile [4]

(b) (i) $q^2 = \frac{48}{160}$

$= 0.3$

$q = 0.55$

but $p + q = 1$

so $p = 1 - 0.55$

$= 0.45$

but $p^2 + 2pq + q^2 = 1$

so $0.45^2 + 2 \times 0.45 \times 0.55 + 0.55^2 = 1$

$0.2 + 0.5 + 0.3 = 1$

BB = 0.2 Bb = 0.5 bb = 0.3 [3]

(ii) BB 2000 Bb 5000 bb 3000 [2]

[Total: 9]

2

A (iv), B (iii), C (v), D, (ii), E (i). [Total: 5]

3

(a) triplet [1]

(b) codes for an amino acid, codes for stop or start [2]

[Total: 3]

4

(a) 8 or 4 pairs [1]

(b) (i) During telophase I of meiosis the chromosomes are bivalent/the centromeres are still intact, whereas in telophase II the chromosomes are single [1]

(ii) During telophase of mitosis the chromosomes are in pairs, whereas in telophase II of meiosis they are single (haploid) [1]

(c) the spindle contracts; pulls the centromeres apart; chromosomes begin to be pulled to both poles. [2]

[Total: 5]

Chapter 5 Variation and selection

1

(a) continuous variation; [1]
(b) two from:
genetic;
the nutrition of the mother;
mother's smoking;
mother's alcohol intake;
mother's health; [2]
(c) three from:
heavy babies have higher death rate;
light babies have higher death rate;
so babies of average mass more likely to survive;
they are more likely to have babies of average mass; [3]
(d) Modern techniques can increase survival of light and heavy babies; they in turn will reproduce; [2]
[Total: 8]

2

(a) **Allopatric speciation** takes place after geographical isolation;
- the rising of sea level splits a population of animals; formerly connected by land creating two islands;
- mutations take place so that two groups result in different species.

Sympatric speciation takes place through genetic variation;
- in the same geographical area;
- mutation may result in reproductive incompatibility;
- perhaps a structure in birds may lead to a different song being produced by the new variant;
- this may lead to the new variant being rejected from the mainstream group;
- breeding may be possible within its own group of variants. [6]

(b) Mate them both with a similar male, to give them a chance to produce fertile offspring.
- If they both produce offspring, take a male and female from the offspring, mate them,
- if they produce fertile offspring then original females **are** from the same species. [2]

[Total: 8]

Chapter 6 Biotechnology and genes

1

(a) Steam sterilisation;
microorganisms cannot enter through air filter;
nutrients are pre-sterilised before entry into fermenter. [3]
(b) Contaminant microorganisms enter the fermenter;
compete with the *Penicillium*; fungus;
penicillin yield reduced. [3]
[Total: 6]

2

(a) Identify the specific section of DNA which contains the gene; this can be done using reverse transcriptase; insert DNA into a vector/insert into *Agrobacterium tumefaciens*; this bacterium/this vector then passes the DNA into the recipient cell. [5]
(b) herbicide kills weeds; which reduces competition; for light or water or minerals; soya plants unharmed [3]
[Total: 8]

3

(a) Beginning of fermentation process shown.
The microorganisms took time to reach maximum production but kept at this level.
Nutrients constantly added. [1]
(b) continuous
product amount reaches a constant level;
nutrients at constant level. [2]
[Total: 3]

Chapter 7 Ecology and populations

1

(a) no significant migration;
no significant births or deaths;
marking does not have an adverse effect. [3]

(b) S = total number of individuals in the total population
S_1 = number captured in sample one, marked and released, i.e. 16
S_2 = total number captured in sample two, i.e. 12
S_3 = total marked individuals captured in sample two, i.e. 5

$$\frac{S}{S_1} = \frac{S_2}{S_3} \quad \text{so, } S = \frac{S_1 \times S_2}{S_3}$$

$S = \frac{16 \times 12}{5}$ Estimated no. of shrews is 38 [2]

(c) Not very reliable because the numbers are quite low. High population numbers are more reliable. [1]

[Total: 6]

2

	Type of behaviour			
	simple reflex	*kinesis*	*positive taxis*	*negative taxis*
A bolus of food reaches the top of our oesophagus and is swallowed.	✓			
Insects move from a cold dry area to a warm humid one.			✓	✓
Springtails (insects) are subjected to increasingly hot conditions, and react by increasing speed in a number of directions. Some go towards the heat source and die.		✓		
A motile alga swims towards light.			✓	

[Total: 5]

3

The opposite sexes recognise each other;
the grebes will only mate with other grebes so are more likely to produce fertile offspring;
mating is synchronised, to coincide with ovulation. [3]

[Total: 3]

4

(a) (i) pioneer or primary coloniser [1]

(ii)
- algae cut off light from plants underneath;
- they die as a result;
- bacteria or fungi or saprobiotics decay them;
- they use a lot of oxygen;
- fish die due to not enough oxygen;
- blood worms increase in number as they are adapted to small amounts of oxygen.

[any 5 points] [5]

(b)
- marginal plants or irises were introduced;
- they spread;
- each year the foliage died and rotted;
- this organic material or humus added to the soil or mud;
- secondary colonisers spread from other areas;
- succession took place.

[any 4 points] [4]

[Total: 10]

Chapter 8 Energy and ecosystems

1

(a) (i) When given fertiliser the grasses competed for resources better that the legumes; some legume species could not grow in these conditions. [2]

(ii) Without fertiliser the grass species did not have enough minerals so did not compete as well; the legumes fixed nitrogen in root nodules so could grow effectively. [2]

(b) Cows grazed on some species more than others/ perhaps trampling by cattle destroyed some species but others were tougher and survived/perhaps waste encouraged the growth of some species whereas others were destroyed. [1]

[Total: 5]

2

(a) Reflected from leaf; passes through leaf; wrong wavelength. [2]

(b) (i) NPP = 8000kJ

(ii) NPP is GPP minus losses from respiration. [2]

(c) Energy is lost at each transfer; Through excretion / egestion / uneaten parts; Not enough energy left for further levels [3]

Notes

Notes

Notes

Index

THE HENLEY COLLEGE LIBRARY